Ionizing and Non-ionizing Radiation

Edited by Otolorin Adelaja Osibote

Published in London, United Kingdom

IntechOpen

Supporting open minds since 2005

Ionizing and Non-ionizing Radiation
http://dx.doi.org/10.5772/intechopen.77474
Edited by Otolorin Adelaja Osibote

Contributors
Chandra Makanjee, Vladimir Kutkov, Hideaki Yamashiro, Young-Nam Kang, Yunji Seol, Jina Kim, Hong Seok Jang, Mario Marcelo Figueroa De La Cruz, Roberto Breslin, Isam Salih, Sarat Dalai, Nikunj U Tandel, Devang Trivedi, Markus Raymond Zehringer, Maher Rizkalla, Felipe Perez, James Rizkalla, Matthew Jeffers, Paul Salama, Cristina Nancy Perez Chumbiauca, Adelaja Osibote

First published in London, United Kingdom, 2020 by IntechOpen
IntechOpen is the global imprint of INTECHOPEN LIMITED, registered in England and Wales, registration number: 11086078, 7th floor, 10 Lower Thames Street, London, EC3R 6AF, United Kingdom
Printed in Croatia

British Library Cataloguing-in-Publication Data
A catalogue record for this book is available from the British Library

Additional hard and PDF copies can be obtained from orders@intechopen.com

Ionizing and Non-ionizing Radiation
Edited by Otolorin Adelaja Osibote
p. cm.
Print ISBN 978-1-78984-141-1
Online ISBN 978-1-78984-142-8
eBook (PDF) ISBN 978-1-83968-440-1

For EU product safety concerns:
IN TECH d.o.o., Prolaz Marije Krucifikse Kozulić 3, 51000 Rijeka, Croatia, info@intechopen.com or visit our website at intechopen.com.

Meet the editor

Otolorin Adelaja Osibote is a senior lecturer at the Department of Mathematics and Physics, Faculty of Applied Sciences, Cape Peninsula University of Technology, South Africa, where she teaches subjects in physics. She graduated with a second class upper in Physics (Hons), an MSc in Engineering Physics (Medical and Health Physics option), and a DSc in Public Health (Medical Physics) from Escola Nacional de Saude Publica, Fiocruz, Brazil. She has authored around 50 papers in international journals and international conference proceedings. She has been involved in a number of research projects funded by her institution. She is a reviewer for several international scientific journals. Her research interest includes irradiation and radioactivity contents of certain foodstuffs, dosimetry studies in diagnostic radiology, radiation protection and safety, and assessment of metal contents in environmental samples.

Contents

Preface

Radiation is known to be present on our planet naturally. Naturally occurring radioactive materials are present in our homes, food and water, and even in the air we breathe. In addition, radiation is present in manmade activities, including X-rays for diagnosis and therapeutic purposes, testing of nuclear weapons, and in nuclear plants for power generation. Depending on the type of radiation, it is important that exposure is minimized to avoid the risk of biological harm to the human body.

This book provides knowledge on the types of radiation, the uses of radiation, the levels of exposure, and radiation protection. I appreciate the efforts of all contributing authors for sharing their knowledge and research findings in this book. The authors are composed of researchers in different fields such as medicine, science, and other areas involved in the use, measurement, and management of radiation.

This book is dedicated to all readers and I hope they will find it useful in their respective areas of research.

The efforts of the service manager and the staff of IntechOpen are highly appreciated. Their professional assistance and cooperation make the function of the editor a walk in the park.

Dr. Otolorin Adelaja Osibote
Department of Mathematics and Physics,
Cape Peninsula University of Technology,
South Africa

Section 1

Radiation Protection and Measurement

Chapter 1

Introductory Chapter: Radiation Exposure, Dose and Protection

Otolorin Adelaja Osibote

1. Introduction

Radiation sources are known to be basically of two origins, that is, the natural or background radiation and artificial or man-made radiation. Natural or background radiation sources are grouped as those from cosmic; these are radiation from the space. The dose from cosmic source of radiation could vary from one location to another, i.e., the dose values vary in different parts of the world and also change with altitude. The exposure also decreases in intensity with depth in the atmosphere or increase with increase altitude [1].

Other natural source group is the terrestrial radiation, which is from soil, water and vegetation. These are radionuclides such as ^{238}U, ^{234}Th and ^{40}K, and they contribute mostly to the external dose to human body. Radon is another example of naturally occurring radionuclide which is found in rock formations and can release higher levels of radiation that can pose health risks particularly lung cancer.

The third source of natural radiation is the internal radiation, and these are ^{40}K, ^{14}C and ^{210}Pb inside the body. These radionuclides enter the body through the ingestion of food, milk and water or by inhalation.

The artificial or man-made radiations are those that originate from various activities of man such as in consumer products, examples which include building materials, television receivers and tobacco products. Other activities are nuclear power plants for electricity/power generation, testing and using of nuclear bombs, decommissioning of radioactive waste, and industrial activities such as mining, security inspection systems use in cargo scanners and personnel security systems and medical purposes.

Also radiation can be categorized into types; they are ionizing and nonionizing radiation. Nonionizing radiation is the type of electromagnetic radiation with no enough energy to ionize atom, while ionizing radiation is radiation that carries enough energy to detach electrons from atoms causing the atom to become charged or ionized. Ionizing radiation has more energy than nonionizing radiation, that is, enough to cause chemical changes, and thereby causing damage to tissue. The ionizing radiation is further categorized into four types: alpha particles, beta particles, gamma rays and X-rays. The effects of ionizing radiation at high-dose levels are well known, while the effects of ionizing radiation at low doses are not yet clear. Ionizing radiation is used for diagnostic and therapeutic medical purposes, and there are advantages and disadvantages attached to the use of ionizing radiation for this purpose; the advantage lies in being able to diagnose and treat diseases; however, it can damage human cells and cause harm. Radiation doses of about 10 Sv and above received in a short period can cause the organs and tissues in the body to cease to function and may lead to death [2].

These two categories of radiation, ionizing and nonionizing, can cause damage to humans. Ionizing radiation can cause cancer, heart and brain problems, while nonionizing radiation can cause burning of retinas, skin cancer as a result of long exposure to the sun [3].

Examples of natural sources of ionizing radiation include metal mining, radon exposure, cosmic rays from the sun and radioactive rocks and soils, while examples of artificial sources of ionizing radiation includes nuclear reactors, medical equipment such as X-rays. Sources of natural nonionizing radiation are sunlight and thermal radiation, while man-made sources of nonionizing radiation are microwave oven, cell phones and power lines.

Most of the man-made exposure to radiation is from medical procedures. This can be shown from the NCRP Report No. 93, 1987, on the ionizing radiation exposure of the population of the United States. Natural sources of radiation accounted for 82%, and medical sources are responsible for 11% of the remaining and 18% from man-made radiation (NCRP Report No. 160), and most of the exposure is from diagnostic X-rays such as examinations of computed tomography, conventional radiography and fluoroscopy and interventional fluoroscopy. The average dose from the use of radiation for treatment purposes is much less than that from diagnostic purposes even though quite a number of exposures may be used in certain treatments such as cancer; only a small number of people are involved, and exposures are limited to small areas where treatment is necessary [4].

Medical use of radiation is known to be the greatest artificial source of doses to human beings at large. Following the improvement in technology and healthcare, this has led to an increase in the usage of radiation; this can be measured by the frequency of procedures and by the levels of individual and collective doses. Medical X-rays are responsible, in Western countries, for at least some 300 man Sv per million inhabitants, representing approximately 90% of man-made source. The common sources of radiation exposure to the population are the natural sources and medical irradiation [5].

The risk of radiation exposure from X-ray such as malignancy, skin damage and cataract is high with increasing number of examination performed. There is an increase in the number of procedures performed and the possibility of more complicated procedures such as interventional procedures that can lead to higher doses to patients and staff. The increasing number of computed tomography (CT) procedures performed also can lead to increase in the collective dose.

Since the diagnostic X-rays take the highest portion of the medical use of radiation or in which human are exposed apart from the natural sources, it is therefore necessary that people or the population are protected which therefore necessitate the need for a radiation protection to be considered in order to eliminate the damage from unnecessary exposure. Even though, the doses from diagnostic radiology are much less than in the treatment of diseases, there is a need to monitor that the dose to the patient is not too low or too high for a particular procedure. According to the International Commission on Radiation Protection (ICRP), radiation protection involves the use of three techniques, and these are justification of practices, optimization of protection and the use of dose limits/levels. Since dose limits do not apply to medical exposure, optimization and justification are therefore important in patients using radiation for medical purposes.

The European Union Council Directive 97/43/Euratom (the Council of the European Union, 1997) also laid emphasis on the need of these two principles of justification and optimization. The principle of justification implies that the advantages to the patient and the society during a radiological procedure must be more than the risks for the patient and the need to consider alternative techniques that do not involve medical radiation exposure [6].

The principle of optimization is to keep the dose 'as low as reasonably achievable' (ALARA principle) economic and social factors being taken into consideration (ICRP 60) [7]. Also ICRP in its recommendation in Publication 73 (ICRP 73) introduced the need for establishment and use of diagnostic reference levels (DRLs) to ensure that implementation guidance is available. The purpose of DRLs is not to be used when considering the dose to individual patients but to prevent delivery of unnecessary high doses as well as to be used in estimating radiation doses as a form of quality assurance [7].

The International Commission on Radiological Protection (ICRP) defined DRL as 'a form of investigation level, applied to an easily measured quantity, usually the absorbed dose in air, or tissue-equivalent material at the surface of a simple phantom or a representative patient,' while the Council of the European Union defined DRL as 'dose levels in medical radiodiagnostic practices or, in case of radiopharmaceuticals, levels of activity, for typical examinations for groups of standard-sized patients or standard phantoms for broadly defined types of equipment.'

DRLs settings for diagnostic radiology should not be based on patient's doses measured from only well-equipped hospitals but in all types of different hospitals, clinics and practices. DRL values are to be established by using the 75th percentile, taking into account of values that are too low or too high. DRLs are to be set locally, regionally or nationally and recorded on regular basis to allow for comparison over some time and also for the purpose of establishing database. According to Vassileva and Rehani [8], DRLs are indicators for a typical practice in a country or in a region, and since equipment and procedures can vary between different facilities in countries or regions, it is therefore a good practice to establish national or regional DRLS. DRLs should be reviewed wherever DRLs are constantly exceeded and that corrective actions are taken when appropriate.

In most countries with established National Diagnostic Reference Levels (nDRLs), the responsibility lies with the government national authorities and institutes responsible for radiological protection and nuclear safety. They perform the function of collecting data from different hospitals or clinics with medical imaging facilities, analysis of the data and then give update on the DRL values. The established DRL values are reviewed periodically, and recommendations are made based on the findings.

2. Reason for DRLs

From the article on historical background on DRLs, Wall and Shrimpton [9] reported that national surveys of patient doses on X-ray examinations conducted in Europe and the USA in the 1950s showed high variations in doses from different hospitals which came about the need for quantitative guidance on patient exposure. It was reported that in the late 1980s, the dose guidelines started first in the USA and then in the UK and then followed in Europe, and the reference doses were incorporated into working documents giving *Quality Criteria for Diagnostic Radiographic Images* for adult and pediatric by European countries study groups of radiologists and physicists.

In 1997, the need to develop the DRLs then followed, (Council Directive 97/43/EURATOM, 1997) which is defined as dose levels in diagnostic radiology to patients of standard-sized groups or standard phantoms, for particular examinations and as well considering different types of equipment [6]. The DRL values should not be exceeded for standard procedures when good and normal practice is applied. The main aim of a DRL is to serve as a control in using radiation for diagnostic purposes and by avoiding unnecessary exposure to radiation. In 1989, national reference doses were first suggested for some radiographic examinations. This was followed by the investigation in the levels in patient doses by ICRP in 1990 and further developed into development of DRLs in ICRP Publication 73.

The list of medical exposure according to the United Kingdom nDRLs required by the Ionizing Radiation Regulations in 2000 include adult and pediatric computer tomography examinations, general radiography and fluoroscopy which include diagnostic examinations on adult and pediatrics and interventional procedures on adult and dental radiography.

3. Regulatory bodies on the use of ionizing radiation and DRLs

The regulatory bodies on the use of radiation include the following organizations:

3.1 United Nations Scientific Committee on the Effects of Atomic Radiation, UNSCEAR

UNSCEAR, which was established in 1955 with the mandate to undertake broad assessments of the sources of ionizing radiation and its effects on human health and the environment, provides the service of assessing global levels and effects of ionizing radiation as well as providing scientific basis for radiation protection. The use of radiation for medical purposes could be of positive applications; it is a reality that X-rays can cause biological harm or injury to humans [10]. Reports from developed countries indicated that the use of ionizing radiation for diagnostic purpose is estimated to be about 1 mSv per capital annual. At this dose level, the estimated annual additional cancer mortality is 0.5 per 10,000 persons of a general population basing on the additive risk model of the United Nations Scientific Committee on the Effect of Atomic Radiation (UNSCEAR). In its report in 2008, UNSCEAR Report No. 1 reported an increase in the total number of diagnostic medical examinations from 2.4 to 3.6 billion; this is an increase of almost 50% from its previous study in 1991–1996. The use of high-dose X-ray techniques such as the computed tomography scanning is leading to growth in the annual number of procedures in many countries thereby increasing the collective dose. It is estimated that the total collective effective dose from medical diagnostic examinations have increased by 1.7 million man Sv, that is, it rises from about 2.3 million to about 4 million man Sv, which gives an increase of about 70% [11].

3.2 International Atomic Energy Agency (IAEA)

IAEA develops safety standards to protect the health and minimize the danger to people's life and property associated with the use of ionizing radiation in medicine, etc. IAEA focuses on ensuring that radiation doses to patients commensurate with the medical purpose, thereby preventing patients from being exposed to unnecessary and unintended radiation. To ensure that radiation protection and safety of radiation sources in medical uses of ionizing radiation, the IAEA Safety Guide on Radiation Protection and Safety in Medical Uses of Ionizing Radiation (2018) was published to provide recommendations and guidance on fulfilling the requirements of IAEA Safety Standards series No GSR Part 3 [12].

According to the report from the IAEA office of Public Information and Communication, DRLs is a tool for comparing diagnostic imaging procedures in a country which include adults and children of different ages and weights in examinations in X-rays, CT, image-guided interventional procedures or nuclear medicine procedure. Each facility needs to set their DRL and then compare with local, national or regional doses. The newsletter report also mentioned the need to track radiation dose data to improve practice and reduce doses without loss of diagnostic quality. As well as prevent unnecessary exposures.

International Atomic Energy Agency also states that DRLs should be set locally, regionally or even nationally. IAEA also agreed to set the nDRLs at the third quartile values, and they could not be considered as optimum dose but in identifying unusual practices. According to IAEA, the government is responsible for the establishment of DRLs and to involve health authority, the professional bodies and the regulatory body. IAEA also identifies DRLs as a tool in radiation protection of the patients.

3.3 The International Commission on Radiological Protection (ICRP)

The primary aim of radiological protection, as stated in ICRP Publication 60, is 'to provide an appropriate standard of protection for mankind without unduly limiting the beneficial practices giving rise to radiation exposure' [13].

According to the International Commission on Radiological Protection (ICRP) in its international recommendations, ICRP 60, (ICRP 19), the focus is on the principles of justification and optimization of all radiation exposures in diagnostic radiology. Another recommendation, which is the ICRP 85, [14], focused on the risk of skin damage from interventional radiology. In 2007 in its publication (ICRP Publication 103), ICRP presented the revised recommendations for radiological protection followed by ICRP Publication 118 (2012) published on deterministic effects of ionizing radiation. ICRP makes recommendations only, and it is the responsibility of government of individual countries to implement those recommendations through legislation appropriate for their own country.

3.4 World Health Organization (WHO)

According to the World Health Organization (WHO), there are established relevant guidelines that have to be considered in each type of diagnostic procedure [15–17]. Human exposure to radiation for medical research is considered as not justified unless it is in accordance with the provisions of the Helsinki Declaration [18] and follows the guidelines for its application prepared by the Council for International Organizations of Medical Sciences [19] and WHO [20].

The WHO in 2008 launched a Global Initiative on Radiation Safety in Health Care Settings (GIRSHCS), thereby facilitating the adoption and applications of regulations, in the evaluation of radiation medicine and medical imaging procedures. WHO also facilitates training on the use of appropriate technologies as well as publishing and disseminating guidance tools and technical documents. In 2012, the WHO presented report of its Radiation Risk Communication in pediatric imaging workshop on the need to develop and implement a risk communication tool in order to create the awareness of radiation risks and exposure in pediatric procedures [21].

3.5 National Council on Radiation Protection and Measurements (NCRP)

National Council on Radiation Protection and Measurements Report No. 160 (1993) focused on the biological effects of ionizing radiation such as cancer, cardiovascular disease and cataracts, while its Report No. 180 focused on the management of exposure to ionizing radiation and expressed radiation protection principles as justification, optimization of protection and numeric protection criteria, i.e., the management of dose to an individual. This means that the protection criteria is the first objective when there is a numeric protection for a specific exposure; then the optimization of protection should follow [22, 23].

4. Conclusion

The use and exposure of humans to ionizing and nonionizing form of radiation is of various purposes. Radiation exposure cannot be entirely avoided on this planet, taking into account how much radiation people receive from natural sources. The proper use of radiation can be of immense benefits. The sources and categories of radiation exposure, the various use of ionizing radiation and the principles of radiation protection to avoid unnecessary exposure to high level of radiation dose from the use of ionizing radiation have been discussed in this chapter.

Author details

Otolorin Adelaja Osibote
Department of Mathematics and Physics, Cape Peninsula University of Technology, Cape Town, South Africa

*Address all correspondence to: osibotea@cput.ac.za

References

[1] United States Environmental Protection Agency. Radiation sources and doses. Available from: https://www.epa.gov/radiation/radiation-sources-and-doses

[2] Australian radiation protection and Nuclear Safety- Health Effects of Ionizing Radiation. Available from: https://www.arpansa.gov.au/understanding-radiation/what-is-radiation/ionising-radiation/health-effects

[3] EMF Academy. Difference Between Ionizing and Non-Ionizing Radiation Written by Christian. 2018. Available from: https://emfacademy.com/difference-ionizing-non-ionizing-radiation/

[4] National Council on Radiation Protection and Measurement NCRP Report No. 93. Ionizing Radiation Exposure of the Population of the United States

[5] Centers for Disease Control and Prevention (CDC). The Electromagnetic Spectrum: Ionizing Radiation. 2015. Available from: https://www.cdc.gov/nceh/radiation/ionizing_radiation.html

[6] European Commission. Council Directive 97/43/EURATOM of 30 June 1997 on health protection of individuals against the danger of ionizing radiation in relation to medical exposure. Official Journal of the European Commission. 180

[7] ICRP. Radiological protection and safety in medicine. ICRP Publication 73. Annals of the ICRP. 1996;**26**(2)

[8] Vassileva J, Rehani M. Diagnostic reference levels. American Journal of Roentgenology 2015;**204**:W1-W3

[9] Wall BF, Shrimpton PC. The historical development of reference doses in diagnostic radiology. Radiation Protection Dosimetry. 1998;**80**(1-3):15-20

[10] United Nations Scientific Committee on the Effects of Atomic Radiation. UNSCEAR 1993 Report to the General Assembly. Sources and Effects of Ionizing Radiation

[11] United Nations Scientific Committee on the Effects of Atomic Radiation. UNSCEAR 2008, Report No. 1. Report to the General Assembly. Sources and Effects of Ionizing Radiation

[12] International Atomic Energy Agency Safety Standards: Radiation Protection and Safety in Medical Uses of Ionizing Radiation. Specific Safety Guide No SSG-48. 2018

[13] ICRP. 1990 Recommendations of the International Commission on Radiological Protection. ICRP Publication 60. Annals of the ICRP. 1991;**21**(1-3)

[14] ICRP. Avoidance of Radiation Injuries from Medical Interventional Procedures. ICRP Publication 85. Annals of the ICRP. 2000;**30**(2)

[15] World Health Organization (WHO). A rational approach to radiodiagnostic investigations. Technical Report Series No. 689. Geneva: WHO; 1983

[16] World Health Organization (WHO). Effective choices for diagnostic imaging in clinical practices. Technical Report Series No. 795. Geneva: WHO; 1990

[17] World Health Organization (WHO). Rational use of diagnostic imaging in paediatrics. Technical Report Series No. 757. Geneva: WHO; 1987

[18] Helsinki Declaration 1964. Adopted by the 18th World Medical Assembly and as amended by the 29th World

Medical Assembly, Tokyo, 1975, the 35th World Medical Assembly, Venice, 1983, and the 41st World Medical Assembly, Hong Kong

[19] Council for International Organizations of Medical Sciences (CIOMS), in collaboration with World Health Organization. International Ethical Guidelines for Biomedical Research Involving Human Subjects. Geneva; 1993

[20] World Health Organization (WHO). Use of ionizing radiation and radionuclides on human beings for medical research, training and non-medical purposes. Technical Report Series No. 611. Geneva: WHO; 1977

[21] World Health Organization. Radiation risk communication in paediatric imaging. Global initiative on radiation safety in health care settings workshop report. 2012

[22] National Council on Radiation Protection (NCRP) Report No. 180—Management of Exposure to Ionizing Radiation: Radiation Protection Guidance for the United States; 2018

[23] NCRP Report No. 160. Ionizing Radiation Exposure of the Population of the United States

Chapter 2

The Effect of Repeated Electromagnetic Fields Stimulation in Biological Systems

Felipe P. Perez, James Rizkalla, Matthew Jeffers, Paul Salama, Cristina N. Perez Chumbiauca and Maher Rizkalla

Abstract

The effects of electromagnetic fields on living organs have been explored with the use of both biological experimentation and computer simulations. In this paper we will examine the effects of the repeated electromagnetic field stimulation (REMFS) on cell cultures, mouse models, and computer simulations for diagnostic purposes. In our biological experiments we used 50 MHz and 64 MHz since this is approved in MRI systems. REMFS upregulated pathways that control the aging process such as proteostasis. REMFS delayed and reversed cellular senescence in mouse and human cell cultures. More recently we determined that REMFS decreases toxic protein beta amyloid levels, which is the cause of Alzheimer's disease (AD), in human neuronal cultures. The mechanism of these effects is the reactivation of the heat shock factor 1 (HSF1). HSF1 activation is a quantum effect of the EMF-oscillations on the water that surrounds a long non-coding RNA, allowing it to then bind and activate the HSF1. We also performed electromagnetic (EM) computer simulations of virtual prototypes of bone cancer, femur fracture, and diabetic foot ulcers utilizing different frequencies and power applications to build an accurate differential diagnosis. These applications indicate the feasibility of subsequent practical models for diagnosing and treating human diseases.

Keywords: electromagnetic, aging, Alzheimer's, simulation, diagnosis, treatment

1. Introduction

Living organisms interact with and adapt to the EMF environment. This discovery has ignited interest in the analysis of the EMF-biological systems [1–4]. Researchers imagine that precise tuning of experimental and clinical REMFS exposure could lead to favorable health results including the development of treatment and diagnostic devices [5, 6]. REMFS exposures produce the activation of multiple biological pathways, including changes in Ca^{2+} regulation [7, 8], channel activity [9], enzyme activity [10], RNA and DNA synthesis [11–13], expression of microRNA [14–16], free radical processes in the genetic effects of EMF [17–19], decreasing oxidative stress [20–24], activation of the heat shock response [25], activation of the heat shock factor 1 (HSF1) [26], cytoprotecting [27], growth behavior [11, 28], activation of the ubiquitin-proteasome [29–31], autophagy-lysosome systems [32], inflammation [33–35], mitochondrial enhancement [36], neuronal activity [37], and a reduction in β-secretase activity [38].

Here, we will focus on studies performed on the REMFS spectrum (50–918 MHz) to explain the mechanism by which non-ionizing, non-thermal, non-modulated, continuous waves cause biological effects. We will use our and other researchers' recent results on human cell cultures and mouse AD models to explain this interaction. Initially, the frequency used in our REMFS experiments was 50 MHz with a specific absorption rate (SAR) of 0.5 W/kg. We found that these exposures upregulated the heat shock factor-1 (HSF1) in human lymphocytes and mouse fibroblasts [39]. HSF1 upregulation increased 70-kDa heat shock protein (HSP70) levels and delayed cellular senescence and death in these cell cultures [39]. Our recent data [40] demonstrated that cultures treated with REMFS at 64 MHz, with a SAR of 0.6 W/kg for 1 h daily for 21 days, had significantly reduced ($p = 0.001$) levels of Aβ40 peptide, compared to untreated cultures [40]. We also demonstrated a quantitative reduction of Aβ levels in primary human neuron cultures after different times and power protocols. There are further therapeutic implications of REMFS based on the improved cognitive function noted by lowering of Aβ levels in several AD mouse model studies performed by other investigators using 918 MHz exposures [36–38, 41].

The aforementioned biological effects demonstrate that REMFS is a multi-target strategy with potentially therapeutic implications on human diseases. In fact, among the biological effects observed, results of our experiments promote the capacity of REMFS to influence various networks of physiological functions that are dysregulated in the aging process and in Late Onset Alzheimer's disease (LOAD) [39, 40, 42].

2. Quantum mechanism of REMFS

The low energy (2–37 eV^{-7}) of the REMFS exposures is not able to cause any chemical change under the provision of classical physics, since the atoms or molecules pass through a potential barrier that they theoretically cannot overcome [43]. Our main challenge is to explain the mechanism of the REMFS-biological system interaction. There is a disparity between the low energy carried by the REMFS perturbation and the response of the biological system. The REMFS biological system interaction is a paradox from the classical point of view as it enables elementary particles and atoms to penetrate an energetic barrier without the need for sufficient energy to overcome it. To solve this paradox we have to look into the quantum scale and examine non-classical physical phenomenon such as wave-particle duality and quantum tunneling [44].

Despite this difficulty here we will describe a plausible sequence of events and a mathematical model for the REMFS-interaction. First we need to consider the REMFS perturbation as a time dependent adiabatic perturbation of water (the most likely receptor for REMFS), specifically the H bond network of the interfacial water [45]. REMFS subjects the quantum system, in the form of the H bond, to gradually changing external conditions giving the quantum system sufficient time for the functional form to adapt [46]. The probabilities of quantum transitions of the adiabatic change in the frequency of the quantum system have been calculated previously on the example of the harmonic oscillator [47].

The reason for the REMFS biological effects rests on the difference between the man-made EMF, and the natural EMF [48]. Man-made EMF waves are produced by parallel EMF oscillating circuits, whereas natural electromagnetic radiation is produced by atomic events such as nuclear fusion from the sun releasing infrared, visible, ultraviolet, X-rays [48]. For this reason, man-made RF-EMF vibrations occur in a single plane, so they are polarized in contrast to the multi plane vibrations from the natural EMF waves. This polarization would explain the differences in the biological effects of man-made versus natural RF-EMF. The polarized RF-EMF exposure has the ability to force all charged/polar molecules and chemical bonds to oscillate on parallel planes, and in phase with the applied polarized field [48, 49]. This external excitatory oscillation forces the exposed physical or chemical system to vibrate at the excitatory frequency changing the frequency of the system to the excitatory frequency [50, 51]. One of the targets of this driven oscillation is the hydrogen bond (HB) network of the first layer of the interfacial water (FLIFW) that surrounds an RNA which in the case of REMFS is a long non-coding RNA (HSR1) [52, 53]. The REMFS oscillations are absorbed by the HB which then acts as a driven quantum harmonic oscillator [54]. This HB responds to REMFS increasing its vibration amplitude [55] with the consequent decreased distance in the direction of the nucleic acid [56]. Since the tunnel probability is proportional to the square of the amplitude, the tunneling probability is increased. REMFS induced quantum tunneling allows proton transfer from the interfacial water to the nucleic acids of RNA [57]. The protonation of the nucleic acid results in tautomeric interconversions [58] with the consequent conformational changes. In the case of the REMFS, a long non-coding RNA called Heat shock RNA (HSR1) changes from a close to an open structure [53] able to bind and activate HSF1 to initiate the expression of several heat shock proteins and chaperones including HSP70 [39].

3. REMFS mathematical model

We developed a mathematical model of the REMFS and biological systems interaction at the quantum level. We hypothesized the quantum effects of REMFS that explain how a low energy exposure is able to produce biochemical changes. For clarity, we divided them into 3 stages with the consequent three equations (see **Figure 1**).

Stage 1. The oscillating REMFS energy causes a time dependent adiabatic perturbation on the first layer of the interfacial water (FLIFW). REMFS perturbs the HB from FLIFW to the oxygen (O) of the "Guanine of the RNA" (GRNA). Under REMFS the H bond of the FLIFW acts as a driven quantum harmonic oscillator increasing the amplitude of the HB vibrations. The following equation estimates the increase in the amplitude of HB vibration when it acts as a driven quantum harmonic oscillator system under REMFS [59]. See following time dependent equation from Piilo et al. to find the amplitude change under REMFS:

$$F(t) = \frac{A\cos(\omega t + \varphi)}{\sqrt{2m\hbar\omega_0}} \tag{1.1}$$

We will obtain the change in the amplitude of the H bond vibration under REMFS:

$$A = \frac{F(t)\sqrt{2m\hbar\omega_0}}{\cos(\omega t + \varphi)} \tag{1.2}$$

Where A describes the amplitude and ω the oscillation frequency of the periodic force, ω_0 is the frequency of the oscillator, m is the mass of the oscillator, φ is the phase of the driving field, t is the time of the exposure, F(t) is the time dependent periodic force, and $\hbar$ is the reduced Planck's constant.

Stage 2. The increased HB vibration amplitude induced by REMFS triggers shortening of the HB of the FLIFW to O of the GRNA. The calculated distance

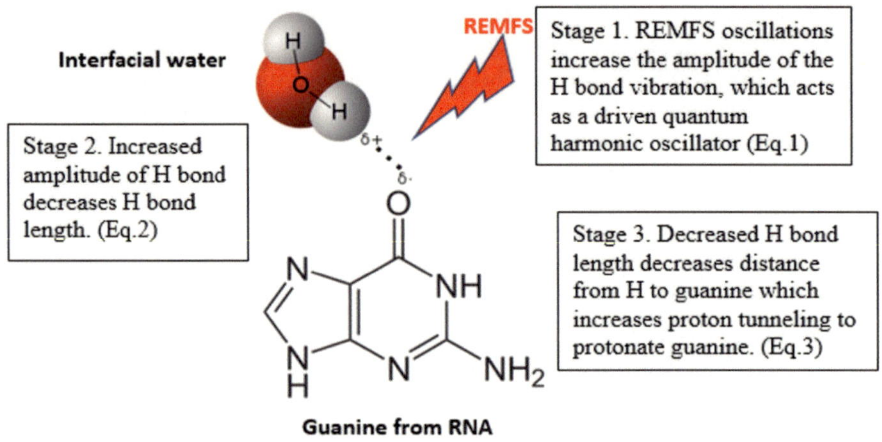

Figure 1.
REMFS quantum effects on the first layer of the interfacial water of the RNA. REMFS (repeated electromagnetic field stimulation).

between the H from the first layer of the interfacial water and the Oxygen of the RNA Oxygen is 1.85 Å [60]. This value is very short, predisposing the H for quantum tunneling. The following equation estimates the change in HB shortening as a function of the amplitude of the oscillation [61].

To find the variation of the HB distance we will use the equation from Samdal et al. using the average over the inter-nuclear configurations of the interfacial water and the RNA nucleic acids (r_a^{exp}) which is given as:

$$r_a^{exp} = \langle r \rangle = \int F_r(q)\, P(q) d\, q \qquad (2.1)$$

$F_r(q)$ = variation of bond distance; P(q) = probability function.

The equation of P(q) in the classical Boltzmann approximation is:

$$P(q) = 1/N \int \exp\left(-\frac{V(q)}{RT}\right) dq \qquad (2.2)$$

This equation predicts the shortening of the HB of the interfacial water under the time dependent perturbation caused by REMFS.

Stage 3. Shortening of the H bond decreases distance from the H from the FLIFW to the O from the GRNA. This will increase the probability of proton tunneling. The following equation estimates the quantum tunneling probability when barrier thickness or distance decreases [43].

$$-\frac{\hbar^2}{2m}\frac{d^2\Psi(x)}{dx^2} = (E - U_0)\Psi(x) \qquad (3)$$

$\Psi = A e^{-\alpha x}$, where $\alpha = \sqrt{\frac{2m(U_0 - E)}{\hbar^2}}$

Where E is a particle of energy, U_0 is the height of the barrier, and $E < U_0$. Also where m is the mass, d is the thickness of the barrier, α is the attenuation factor.

These three equations formed the mathematical model that is able to predict how the time dependent perturbation caused by REMFS affects the HB network of the first layer of the interfacial water. This HB acts as a quantum harmonic oscillator to produce proton tunneling and the protonation of the nucleic acids of the surrounding RNA to produce biological effects.

4. REMFS biological effects

4.1 REMFS effects in aging and age-related diseases

Multiple studies have found that repeated mild heat-shock (RMHS) produce beneficial effects on human fibroblasts experiencing in vitro senescence [62]. This prompted our laboratory to study the effects of a similar electromagnetic stimuli but instead applying lower frequency energy and therefore less risk of producing heat damage (430 THz vs. 64 MHz). The fact that the energy of the REMFS exposures are several orders of magnitude lower energy that heat exposure, yet both heat and REMFS are able to activate the HSR, makes REMFS a more suitable and safer strategy to activate this biological pathway to prevent and treat age-related diseases. We hypothesized that these exposures would produce anti-aging changes such as delay of in vitro senescence, would also lead to retardation of progressive cell enlargement, prevention of development of abnormal proteins, increased glutathione, and decrease in age-dependent glycosylation [63], as well as maintenance of youthful morphology, increased proteasome activity, increased levels of various heat shock proteins (HSP's), increased resistance against oxidative abilities, and UV-A irradiation similar to repeated mild heat shock [64]. Interestingly many of these anti-aging effects are produce by the heat shock response (HSR) elements [65]. In fact, attenuation in the HSR during senescence is the earliest event in the aging process, and is characterized by loss of proteostasis [66] that comes as a result of decreased heat shock factor-1 (HSF1) DNA-binding [67].

4.1.1 REMFS delays cellular senescence

Originally our laboratory utilized a frequency of 50 MHz, a power of 0.5 W, and a specific absorption rate (SAR) of 0.6 W/kg to expose different types of cell cultures applying different dose regimes [39]. REMFS treated cell cultures showed anti-aging effects. The proposed mechanism is the activation of HSF1 when REMFS releases HSF1 from its repressor Hsp90 to activate it. This study suggests that EMF exposure directly interacts with the HSF1-Hsp90 complex, releasing HSF1 for activation preparing it for following injuries. Our experiments also revealed that REMFS increased the population doublings number and changed the morphology of the cells to youthful appearance near the end of their replicative life in wild type, but not in the knockout HSF1 cell cultures. REMSF also decreased cell mortality in human T-cells. Remarkably, REMFS also increased HSF1 phosphorylation, enhanced HSF1-DNA binding, and improved HSP70 expression relative to non-treated cells [39].

We hypothesized a mechanism in which REMFS oscillation produces increase amplitude of the hydrogen bond of the interfacial water, therefore increasing the probability of proton tunneling. This proton transfer between the hydrogen bond of the interfacial water and the oxygen of the adjacent nucleic acids of the heat-shock-RNA1 (HSR1) will protonate the nucleic acid to form tautomers [58] that will cause conformational changes in this long non-coding RNA [68]. This secondary structure would be able to bind HSF1 and activate it by dissociation from the repressor chaperone HSP90 [39]. Then, activated HSF1 enters the nucleus and binds DNA to induce expression of beneficial chaperones and, ultimately, the promotion of anti-aging and proteostasis effects [69]. The REMSF exposure utilized here is a potential new strategy to treat age-related diseases such as Alzheimer's. We will examine the experiments from REMFS exposures in human neuronal cultures [40] and the studies from other investigators in AD mouse models [41].

4.1.2 REMFS in human neurons

One of the hallmarks of the aging process is the decrease of the proteostasis due to attenuation of the HSF1 which produce protein aggregation [42]. Alzheimer's is the most common protein deposition disease, it is caused by beta amyloid (Aβ) aggregation. Our recent study showed that REMFS decreased Aβ levels in human neuronal cultures [40]. REMFS decreased Aβ 1–40 and Aβ 1–42 levels. Importantly, it did not cause any toxicity in the neuronal cultures. We tested several REMFS parameters such as time of exposure, frequency, etc. to define any change in the levels of Aβ. Initially we used REMFS treatments at 64 MHz with a SAR of 0.6 W/kg daily for 1 h for 21 days. Results showed a decrease of 58.3% in Aβ 1–40 levels. We also found that these treatments did not cause any toxicity to the cultures compared to control, non-treated cells as measured by LDH levels, cell morphology, cell attachment, cell number, or neurite extension. Subsequently, we decided to determine if 14 days of REMFS at 64 or 100 MHz with a lower SAR of 0.4 W/kg also decreased the Aβ 1–42 levels. We found that there was a similar significant difference in the Aβ 1–42 levels when we increased the exposure time from 1 to 2 h or when we put the chamber outside the incubator. When we increased the frequency from 64 to 100 MHz; we found a similar beneficial effect in Aβ 1–42 levels. This suggest that REMFS at 64 MHz with a SAR of 0.4 W/kg for 1 h could be the minimal energy required to induce reduction of the Aβ peptides levels, these results are important for future clinical trials. All these suggest that the decrease of Aβ levels in cell cultures were mediated by the activation of the proteostasis master regulator HSF1 [39], this activated HSF1 would increase the expression of chaperones to induce Aβ degradation.

4.1.3 REMFS in AD mouse models

The first REMFS study that prevented cognitive impairment in a transgenic (Tg) AD mouse model "Transgene with human APP gene bearing the Swedish mutation" (AβPPsw) was performed with a pulsed/modulated RF-EMF at 918 MHz which produced a SAR of 0.25–1.05 W/kg over a 7–9 month period [41]. Non-treated Tg mice showed cognitive impairment in memory tests, on the other hand treated Tg mice preserved memory after 6–7 months of REMFS treatment. A more recent REMFS experiment applying daily exposures for a two-month period in older Tg mice (21–27 months) showed improvement in the Y-maze task (memory test), though did not show improvement in more complex tests after 2 months of REMFS [36]. Also the old non-treated Tg had very high levels of Aβ deposition in most areas of the brain. Conversely, the EMF-treated Tg mice exhibited an impressive 24–30% removal of Aβ deposits, suggesting a disaggregation of pre-existing Aβ deposits following 2 months of daily EMF treatment. More importantly, these long-term (daily for up to 9 months) exposure schedules were found to be very safe because they did not demonstrate any harmful effects including, no effects on brain oxidative stress or abnormal brain histology, no significant brain heating, no damage to DNA in circulating blood cells, and no gross changes to peripheral tissues. Another study performed at a higher frequency (1950 MHz, SAR 5 W/kg, 2 h/day, 5 days/week) was also found to ameliorate AD pathology in Tg-5xFAD and wild type (WT) mice exposed to REMFS for 8 months [38]. Remarkably, long-term REMFS significantly decreased not only Aβ plaques, APP, and APP carboxyl-terminal fragments in whole brain (including hippocampus and entorhinal cortex), but also inhibited the parenchymal expression of beta-amyloid precursor protein cleaving enzyme 1 (BACE1) and neuro-inflammation. Additionally, REMFS recovered memory impairment in

AD mice. Furthermore, treated Tg showed expression of 5 genes (Tshz2, Gm12695, St 3 gal1, Isx and Tll1), which are associated with Aβ metabolism. We found that these genes are significantly altered inTg-5xFAD mice, showing diverse responses to the treatments in the hippocampus of wild control and transgenic mice. Treatment in wild type mice showed no difference than control Tg. Conversely, REMFS-treated Tg group showed contrasting gene expression arrays. All these findings suggest REMFS treatments positively alter Aβ deposition and metabolism in AD, but not in wild type mice [38].

Together, human neurons and AD model mouse experiments suggest that REMFS exposures decrease Aβ at the extra and intra cellular levels. Different from the clinical trials with active and passive immunization, REMFS did not cause encephalitis or inflammation. REMFS has important effects in preventing and decreasing brain Aβ deposition, therefore making REMFS a potential therapeutic strategy in the treatment of advanced AD patients who have massive Aβ aggregation in the brain.

4.1.4 REMFS potential therapeutic strategy

Considering the REMFS effects in Tg AD mice, the results on primary neuronal cultures are very promising as the REMFS parameters such as frequency and SAR we applied creates an appropriate and safe potential new therapeutic strategy for human exposures. However, before we exposed humans to this type of RF radiation, we need to recognize that extrapolating effects of mice exposure to effects of human exposure is complex. The mouse's geometry, size, tissue penetration, tissue dielectric properties are significantly different from that of a human and therefore the external fields produced during the 915 MHz exposure would result in quite different internal fields. Internal fields are the electromagnetic fields inside the object, and not the electromagnetic fields incident upon the object. The energy absorbed by an object is directly related to its internal field. Consequently, it is imperative to determine what type of external fields could yield the same internal fields in mouse and human. An important EMF parameter to contemplate is tissue penetration; we should consider that tissue penetration is inversely proportional to the EMF frequency. For example, 918 MHz (frequency used in mouse experiments and cell phone) has a skin penetration of less than 4 cm in a human head, while it is a whole-body exposure for a mouse. Human head thickness for an adult male is around 19.4 cm, so to irradiate the center of a human head the exposure should have a skin penetration of minimum of 9.7 cm. This demonstrates that the 918 MHz frequency is not able to reach important deep brain areas such as the hippocampus. Additionally, 918 MHz produced a greater energy than REMFS, so rising the thermal injury risk. Instead, REMFS exposure (64 MHz) has a skin penetration of 13.49 cm, similar to the 14.5 width of a human head making it suitable for a human head.

Using similarities in dosimetry between cell cultures, animal exposure, human phantom exposure, and computer simulation it is possible to adjust conditions for human exposure [70]. Thus, we used frequencies more suitable for human exposures (50–75 MHz). The basis for these frequencies was:

1. MRI machines has been used 64 MHz for several decades giving a safe exposure that is similar to the 50 and 64 MHz used in our previous experiments [39].

2. It is similar to the human whole body resonant frequency (75 MHz), [71] at this frequency the body absorbs up to 10 times as much as power as when it is not in

resonance [72]. Consequently, we would need to apply less power and achieve the minimum SAR that could achieve biological effect, a safer exposure compared to high energy fields. This would decrease the complexity of the EMF-biological system interaction decreasing the heat production from the exposure.

The physical and biological conditions of the exposed target would affect the EMF parameters of the exposures concerning the case under study [73].

3. Our REMFS exposures produced a SAR (0.4–0.9 W/kg) well below the value limits values of 2 W/kg set by the Institute of Electrical and Electronics Engineers (IEEE) [40], so offering a safe framework for clinical trials [39]. The REMFS parameter for human exposures will range from daily to twice weekly, with a length extending from 30 min to 1 h for several months founded on human neurons and AD mouse studies [36, 37, 39–41].

4.1.4.1 REMFS anti-aging effects

In our previous studies, we determined that REMFS enhanced the HSF1-DNA and delayed the aging process. Taking into consideration that the decline in proteostasis is the earliest event in the aging process and that it is caused by attenuation of the HSF1-DNA binding [66, 74], this makes REMFS a potential therapeutic strategy to treat age-related diseases.

Nevertheless, we should take into consideration other pathway imbalances that cause the pathomolecular mechanism of age-related diseases. Take, for example, the Forehead box protein (FoxO) pathways whose dysregulation results in accelerating the aging process [75]. This suggests that delaying the aging process may be achieved by reactivation of both HSF1 and FoxO pathways (longevity pathways). The combination of the treatments for these two pathways such as HSF1 enhancers (REMFS) in combination with caloric restriction mimetics such as resveratrol (RV) would be an appropriate therapeutic strategy [69, 76]. Enhancing these two pathways that control an array of different processes, including metabolism, cognition, stress response, and brain plasticity demands close monitoring to prevent hyperstimulation of either pathway, thus controlling side effects [69]. For this reason, we suggest using REMFS because 64 MHz affords a safe framework for human treatments [77]. Our previous studies utilized a non-ionizing EMF radiation of 50 MHz allowed safe exposures comparable to our recent study in human neurons with 64 MHz [39]. We should take into consideration that 918 MHz has less skin penetration and therefore the energy carried by the exposure is absorbed adjacent to the skin. In an interest study the RF exposure during 30 min with a 2.7–4 W/kg SAR, versus a 16 min with 6 W/kg both caused a noteworthy temperature change (0.1–0.4°C), as well as other physiological changes in heart rate, localized sweating, and blood flow [78], thus, we suggest lower and effective SAR values (0.4–0.9 W/kg) to prevent these side effects. REMFS can be applied through different exposure systems such as antennas in anechoic chambers or large TEM chambers, these chambers would likely about 10 m in length, 6 m in height, and 6 m in width and utilized frequencies between 50 and 64 MHz [79, 80]. The Institute of Electrical and Electronics Engineers (IEEE) recommend maximum permissible exposure (MPE) values of less than 1 W/kg [81]. Our REMFS treatment produce SAR's under this limit, so suggesting that this is a safe exposure for human treatments. An important aspect to consider is the homocysteinylation of the HSF1 which could be the cause of the age-related attenuation of the HSF1-DNA binding. Therefore, decreasing plasma homocysteine levels by dietary interventions is recommended to prevent the HSF1-DNA binding [82].

Likewise, FoxO activation is a very crucial part of the combination therapy to delay the aging process and age-related diseases. RV is one of the most effective FoxO activators; it has few side effects and it is easy to administer. RV also activates the mTOR, and SK1N pathways [83]. RV has effects on multiple pathways such as antioxidant, vasodilating, inflammation, cell growth, atherosclerosis, anticoagulant, and beneficial for the cardiac rhythm. Notably, RV decreases mortality and metabolic syndrome in high-calorie and high-fat diets in mice experiments [84]. For these reasons RV is a potentially a new therapeutic strategy to prevent and treat metabolic syndrome and diabetes mellitus type II. One disadvantage is that RV bioavailability is poor as a consequence of metabolic alterations in the plasma. Hypothetically, REMFS combines with RV as soon as a decline in any of these two pathways is detected. One of the methods to determine if the HSF1 pathway is failing would be monitoring the T lymphocyte HSF-1 DNA binding [69]. The method to detect a decline in the FoxO pathway includes testing the FoxO3a binding to DNA. An important part of the evaluation is to determine the aggregation of beta amyloid (Aβ), Tau, or α-synuclein proteins in the human brain using Positron-emission tomography (PET) scanning to monitor neuro-degeneration and protein deposition load [69, 85].

4.1.4.2 REMFS in Alzheimer's disease and other protein aggregation diseases

While REMFS might affect the organism in a whole-body basis, we also consider that more focused exposures, individual body targets may be selected. Any organ that shows functional decline, including the brain, kidneys, joints, liver, or heart, may benefit from engineered REMFS to induce protein disaggregation by activation of the HSF1 pathway. Therefore, we will initiate human head exposure to treat the most common cause of dementia (Alzheimer's disease). Before clinical trials are considered we have to determine the best electromagnetic settings for human exposures such as power output, power deposition, far field, antenna type, distance from antenna, electric field, magnetic field, etc. that will produce uniform internal fields similar to our previous studies when applied to a human brain with a target SAR of 0.4–0.9 W/kg [40]. Initially, we determined by mathematical and computer modeling that the REMFS exposures in our biological studies delivered a safe thermal and SAR measurements [70]. With these results we developed a virtual exposure system by numerical model and computer simulation. We designed a virtual antenna that delivers a SAR of around 0.6 W/kg to a simulated phantom of a human brain. With these simulations we found the REMFS parameters that would deliver a uniform radiation to a human skull in clinical trials [86]. In the near future, we will experimentally confirm these results using an appropriate antenna to expose a Specific Anthropomorphic Mannequin (SAM) human head phantom [87] with internal and external probes oriented vertically to determine the EMF parameters that will provide an effective and safe SAR for future Alzheimer's treatment. Data suggest that the ideal environment for these treatments should be an anechoic chamber to prevent RF wave reflections and provide a uniform exposure to the subjects. The final step will be to initiate phase 1 clinical trials in patient with early Alzheimer's disease to determine safety and efficacy of this new potential therapeutic strategy.

5. Future diagnostic procedures

We performed several computer modeling and simulation to create visual representations of the interior of the human body for diagnostic analysis, as well as visual representation of the function of some organs or tissues. We utilized

EMF of different frequencies up to 5 GHz because they are commonly used in medicine for diagnosis. Here, we show several future non-invasive EMF diagnostic procedures.

5.1 Pathological bone

We performed microwave and thermal simulation of human bone. The results showed differential power dissipation over the bone materials with different temperatures within 2–4° change for various frequencies [88]. This simulation also showed the distinction between normal and abnormal bone tissues, indicating that this is an effective method for diagnosing normal bone and pathological bone including bone cancer, fractures and infection.

5.2 Femoral neck vasculature

We also performed simulations of the vasculature of the femoral bone [89]. Disruption of the blood supply to the femoral neck is a well-recognized source of morbidity and mortality, often resulting in avascular necrosis of the femoral head. EM simulations of femoral neck fractures were presented as examples. Electric fields were generated in a fashion that exploited disruptions within the vasculature of the femoral neck. Simulated blood vessels were developed in two-dimensions: the phi direction (the circular), and the z-direction. Two different frequencies, 3 and 5 GHz were considered, with 100-J energy pulses within blood vessels of 2.54 mm in diameter. The fat surrounding the bone was simulated, we also developed an additional model with layered fat and skin above the vessels. We were able to visualize the femoral neck's blood vessels. This research validated the technique of detecting and diagnosing pathology of the circulation of femur bone in humans. The approach using the characteristics of the RF response of the reflected power at various frequencies as determined from the finite element simulation was appropriate, and it fits well with the practical model if implemented via MEMS (micro-electro-mechanical systems). Magnetic sensors may be built on flexible substrates in order to shape up the sensors and make them suitable for measuring various sizes. The COMSOL models were made close to the anatomical model seen in **Figure 2**. It shows the head, neck, and leg of the femur. The exploitation of electric field indicates the feasibility of a subsequent practical model to diagnose femur vasculature pathology including avascular necrosis of the femoral neck and other human bones.

5.3 Arteriosclerosis disease

We lastly performed simulations to detect arteriosclerosis of human blood vessels which is associated with coronary artery and peripheral vascular disease. Our laboratory developed a new non-invasive EMF approach for the diagnosis of stenosis/arteriosclerosis disease. A simulated human foot was analyzed using COMSOL multi-physics software in attempt to visualize, analyze, and quantify the degree of peripheral vascular disease, which plays a pivotal role in the development of diabetic foot ulcers. The simulation results served as a proof of concept for predicting and stratifying certain degrees of occlusion within the peripheral vasculature. Although this study was based on computer modeling with simulation results in nature, the research parameters shows promise for practical models for future diagnosis of the peripheral vasculature via EM parameters. The study shows promises for the practical implementation of the device. Current technologies with MEMS/NEMS can serve as hardware systems

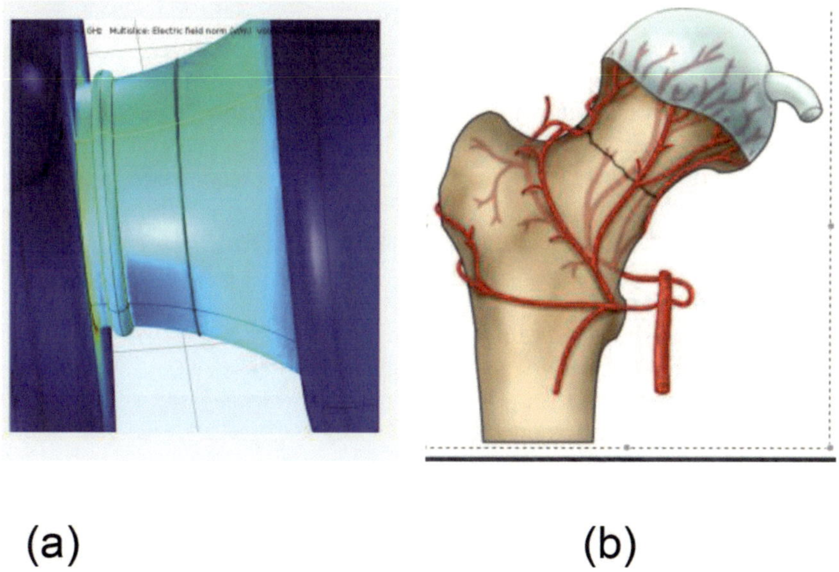

Figure 2.
(a) The computerized model of the femur (b) The anatomical femur with components and blood vessels: blood vessels showing rupture in the femur vasculature.

proper for this diagnosis process designed for detecting EM parameters needed for the diagnostic tool for the early detection of peripheral vascular disease, and ultimately, diabetic foot ulcers [90].

6. Conclusion

Since the discovery of electromagnetic fields, the beneficial health effects and their potential applications toward the treatment and diagnostic of age-related diseases has been eagerly sought with promising results. The effect of non-thermal, non-ionizing REMFS has been examined in our laboratory for its ability to induce cytoprotecting effect via the heat shock factor-1. Results suggest anti-aging effects occurred as a direct consequence of a biological systems-REMFS interaction, and herein we have proposed a quantum tunneling-based mechanism mediated by the interfacial water to explain it. Our pioneering studies have also demonstrated safe REMFS decreases toxic Aβ levels in primary human brain cell cultures; an outcome likely resulting from increased Aβ degradation. When considered in parallel with several transgenic AD mouse model studies that have demonstrated the efficacy and safety of REMFS in-vivo to induce removal/disaggregation of pre-existing Aβ deposits and prevent or reverse cognitive impairment, the potential application of REMFS toward the treatment of AD and age-related protein deposition diseases is certainly encouraging. Furthermore, the simultaneous modulation of longevity pathways through HSF1 enhancers (e.g., REMFS) and FoxO pathway up-regulators (caloric restriction mimetics, such as resveratrol) suggest complementary strategies could act synergistically to balance and preserve cellular defense and repair systems. As REMFS targets the most important pathways affected in Alzheimer's disease and other age-related pathologies, HSF1 modulation and enhancement by REMFS could potentially restore a variety of damaged signaling networks associated with the aging process, additionally, diagnostic EMF devices could prove to be a fast, non-invasive, and painless tool that will avoid incisions into the body and the removal of tissue for diagnosis of a multitude of diseases.

Acronyms

EMF	electromagnetic fields
REMFS	repeated electromagnetic fields stimulation
SAR	specific absorption rate
HSF1	heat shock factor 1
HB	hydrogen bond
FLIFW	first layer of the interfacial water
GRNA	guanine of the RNA
O	oxygen
HSP	heat shock proteins
HSR1	heat shock RNA
RV	resveratrol
FoxO	Forkhead box protein
Aβ	amyloid beta
AD	Alzheimer's disease
SOD	superoxide dismutase
WT	wild type
Tg	transgenic

Author details

Felipe P. Perez[1], James Rizkalla[2], Matthew Jeffers[3], Paul Salama[3], Cristina N. Perez Chumbiauca[1] and Maher Rizkalla[3,4*]

1 Indiana University School of Medicine, Indianapolis, Indiana, United States

2 Baylor University Medical Center, Dallas, Texas, United States

3 Department of Electrical and Computer Engineering, Indiana University Purdue University Indianapolis (IUPUI), Indiana, United States

4 Integrated Nanotechnology Development Institute (INDI), IUPUI, Indianapolis, Indiana, United States

*Address all correspondence to: mrizkall@iu.edu

References

[1] Rosen A, Stuchly MA, Vorst AV. Applications of RF/microwaves in medicine. IEEE Transactions on Microwave Theory and Techniques. 2002;**50**(3):963-974

[2] Michaelson SM. Health implications of exposure to radiofrequency/microwave energies. British Journal of Industrial Medicine. 1982;**39**(2):105-119

[3] Singh S, Kapoor N. Health implications of electromagnetic fields, mechanisms of action, and research needs. Advances in Biology. 2014;**2014**:24

[4] Hardell L, Sage C. Biological effects from electromagnetic field exposure and public exposure standards. Biomedicine & Pharmacotherapy = Biomedecine & Pharmacotherapie. 2008;**62**(2):104-109

[5] Gherardini L, Ciuti G, Tognarelli S, Cinti C. Searching for the perfect wave: The effect of radiofrequency electromagnetic fields on cells. International Journal of Molecular Sciences. 2014;**15**(4):5366-5387

[6] Perez FP, Morisaki JJ, Bandeira JP. Repeated electromagnetic field stimulation in aging and health. In: The Science of Hormesis in Health and Longevity. Cambridge, Massachusetts, USA: Elsevier; 2019. pp. 189-197

[7] Conti P, Gigante GE, Alesse E, Cifone MG, Fieschi C, Reale M, et al. A role for Ca^{2+} in the effect of very low frequency electromagnetic field on the blastogenesis of human lymphocytes. FEBS Letters. 1985;**181**(1):28-32

[8] Rozek RJ, Sherman ML, Liboff AR, McLeod BR, Smith SD. Nifedipine is an antagonist to cyclotron resonance enhancement of 45Ca incorporation in human lymphocytes. Cell Calcium. 1987;**8**(6):413-427

[9] Fesenko EE, Geletyuk VI, Kazachenko VN, Chemeris NK. Preliminary microwave irradiation of water solutions changes their channel-modifying activity. FEBS Letters. 1995;**366**(1):49-52

[10] Byus CV, Pieper SE, Adey WR. The effects of low-energy 60-Hz environmental electromagnetic fields upon the growth-related enzyme ornithine decarboxylase. Carcinogenesis. 1987;**8**(10):1385-1389

[11] Goodman R, Henderson AS. In: SpringerLink, editor. Mechanistic Approaches to Interactions of Electric and Electromagnetic Fields with Living Systems. New York: Plenum Press; 1987

[12] Liboff AR, Williams T Jr, Strong DM, Wistar R Jr. Time-varying magnetic fields: Effect on DNA synthesis. Science. 1984;**223**(4638):818-820

[13] Grundler W, Abmayr W. Differential inactivation analysis of diploid yeast exposed to radiation of various LET. I. Computerized single-cell observation and preliminary application to X-ray-treated *Saccharomyces cerevisiae*. Radiation Research. 1983;**94**(3):464-479

[14] Capelli E, Torrisi F, Venturini L, Granato M, Fassina L, Lupo GFD, et al. Low-frequency pulsed electromagnetic field is able to modulate miRNAs in an experimental cell model of Alzheimer's disease. Journal of Healthcare Engineering. 2017;**2017**:10

[15] Liu Y, Liu WB, Liu KJ, Ao L, Cao J, Zhong JL, et al. Extremely low-frequency electromagnetic fields affect the miRNA-mediated regulation of signaling pathways in the GC-2 cell line. PLoS One. 2015;**10**(10):e0139949

[16] Dasdag S, Akdag MZ, Erdal ME, Erdal N, Ay OI, Ay ME, et al. Effects

of 2.4 GHz radiofrequency radiation emitted from Wi-Fi equipment on microRNA expression in brain tissue. International Journal of Radiation Biology. 2015;**91**(7):555-561

[17] Luukkonen J, Liimatainen A, Juutilainen J, Naarala J. Induction of genomic instability, oxidative processes, and mitochondrial activity by 50Hz magnetic fields in human SH-SY5Y neuroblastoma cells. Mutation Research. 2014;**760**:33-41

[18] Jouni FJ, Abdolmaleki P, Ghanati F. Oxidative stress in broad bean (*Vicia faba* L.) induced by static magnetic field under natural radioactivity. Mutation Research. 2012;**741**(1-2):116-121

[19] Campisi A, Gulino M, Acquaviva R, Bellia P, Raciti G, Grasso R, et al. Reactive oxygen species levels and DNA fragmentation on astrocytes in primary culture after acute exposure to low intensity microwave electromagnetic field. Neuroscience Letters. 2010;**473**(1):52-55

[20] Hajnorouzi A, Vaezzadeh M, Ghanati F, Jamnezhad H, Nahidian B. Growth promotion and a decrease of oxidative stress in maize seedlings by a combination of geomagnetic and weak electromagnetic fields. Journal of Plant Physiology. 2011;**168**(10):1123-1128

[21] Maaroufi K, Had-Aissouni L, Melon C, Sakly M, Abdelmelek H, Poucet B, et al. Spatial learning, monoamines and oxidative stress in rats exposed to 900 MHz electromagnetic field in combination with iron overload. Behavioural Brain Research. 2014;**258**:80-89

[22] Osera C, Amadio M, Falone S, Fassina L, Magenes G, Amicarelli F, et al. Pre-exposure of neuroblastoma cell line to pulsed electromagnetic field prevents H_2O_2-induced ROS production by increasing MnSOD activity. Bioelectromagnetics. 2015;**36**(3):219-232

[23] Di Carlo AL, White NC, Litovitz TA. Mechanical and electromagnetic induction of protection against oxidative stress. Bioelectrochemistry. 2001;**53**(1):87-95

[24] Osera C, Fassina L, Amadio M, Venturini L, Buoso E, Magenes G, et al. Cytoprotective response induced by electromagnetic stimulation on SH-SY5Y human neuroblastoma cell line. Tissue Engineering Parts A. 2011;**17**(19-20):2573-2582

[25] Goodman R, Blank M. Magnetic field stress induces expression of hsp70. Cell Stress & Chaperones. 1998;**3**(2):79-88

[26] Lin H, Opler M, Head M, Blank M, Goodman R. Electromagnetic field exposure induces rapid, transitory heat shock factor activation in human cells. Journal of Cellular Biochemistry. 1997;**66**(4):482-488

[27] Carmody S, Wu XL, Lin H, Blank M, Skopicki H, Goodman R. Cytoprotection by electromagnetic field-induced hsp70: A model for clinical application. Journal of Cellular Biochemistry. 2000;**79**(3):453-459

[28] Blank M, Goodman R. DNA is a fractal antenna in electromagnetic fields. International Journal of Radiation Biology. 2011;**87**(4):409-415

[29] Eleuteri AM, Amici M, Bonfili L, Cecarini V, Cuccioloni M, Grimaldi S, et al. 50 Hz extremely low frequency electromagnetic fields enhance protein carbonyl groups content in cancer cells: Effects on proteasomal systems. Journal of Biomedicine and Biotechnology. 2009;**2009**:834239

[30] Caraglia M, Marra M, Mancinelli F, D'Ambrosio G, Massa R, Giordano A, et al. Electromagnetic fields at mobile

phone frequency induce apoptosis and inactivation of the multi-chaperone complex in human epidermoid cancer cells. Journal of Cellular Physiology. 2005;**204**(2):539-548

[31] Hirai T, Taniura H, Goto Y, Ogura M, Sng JC, Yoneda Y. Stimulation of ubiquitin-proteasome pathway through the expression of amidohydrolase for N-terminal asparagine (Ntan1) in cultured rat hippocampal neurons exposed to static magnetism. Journal of Neurochemistry. 2006;**96**(6):1519-1530

[32] Marchesi N, Osera C, Fassina L, Amadio M, Angeletti F, Morini M, et al. Autophagy is modulated in human neuroblastoma cells through direct exposition to low frequency electromagnetic fields. Journal of Cellular Physiology. 2014;**229**(11):1776-1786

[33] Pena-Philippides JC, Yang Y, Bragina O, Hagberg S, Nemoto E, Roitbak T. Effect of pulsed electromagnetic field (PEMF) on infarct size and inflammation after cerebral ischemia in mice. Translational Stroke Research. 2014;**5**(4):491-500

[34] Kubat NJ, Moffett J, Fray LM. Effect of pulsed electromagnetic field treatment on programmed resolution of inflammation pathway markers in human cells in culture. Journal of Inflammation Research. 2015;**8**:59-69

[35] Rohde CH, Taylor EM, Alonso A, Ascherman JA, Hardy KL, Pilla AA. Pulsed electromagnetic fields reduce postoperative interleukin-1beta, pain, and inflammation: A double-blind, placebo-controlled study in TRAM flap breast reconstruction patients. Plastic and Reconstructive Surgery. 2015;**135**(5):808e-817e

[36] Arendash GW, Mori T, Dorsey M, Gonzalez R, Tajiri N, Borlongan C. Electromagnetic treatment to old Alzheimer's mice reverses beta-amyloid deposition, modifies cerebral blood flow, and provides selected cognitive benefit. PLoS One. 2012;**7**(4):e35751

[37] Arendash GW. Transcranial electromagnetic treatment against Alzheimer's disease: Why it has the potential to trump Alzheimer's disease drug development. Journal of Alzheimer's Disease: JAD. 2012;**32**(2):243-266

[38] Jeong YJ, Kang GY, Kwon JH, Choi HD, Pack JK, Kim N, et al. 1950 MHz electromagnetic fields ameliorate Abeta pathology in Alzheimer's disease mice. Current Alzheimer Research. 2015;**12**(5):481-492

[39] Perez FP, Zhou X, Morisaki J, Jurivich D. Electromagnetic field therapy delays cellular senescence and death by enhancement of the heat shock response. Experimental Gerontology. 2008;**43**(4):307-316

[40] Perez FP, Bandeira JP, Bailey JA, Morisaki JJ, Lahiri DK. Potential noninvasive approach for Alzheimer's disease: Repeated electromagnetic field stimulation lowers beta-amyloid protein levels in primary human neuronal cultures. Journal of the American Geriatrics Society. 2017;**65**(S1):S119-SS20

[41] Arendash GW, Sanchez-Ramos J, Mori T, Mamcarz M, Lin X, Runfeldt M, et al. Electromagnetic field treatment protects against and reverses cognitive impairment in Alzheimer's disease mice. Journal of Alzheimer's Disease: JAD. 2010;**19**(1):191-210

[42] Perez FP, Bose D, Maloney B, Nho K, Shah K, Lahiri DK. Late-onset Alzheimer's disease, heating up and foxed by several proteins: Pathomolecular effects of the aging process. Journal of Alzheimer's Disease: JAD. 2014;**40**(1):1-17

[43] Ankerhold J. Introduction. In: Quantum Tunneling in Complex Systems. Berlin, Germany: Springer; 2007. pp. 1-5

[44] Schmitz KS. Physical Chemistry: Concepts and Theory. Cambridge, Massachusetts, USA: Elsevier; 2016

[45] Lobyshev V. Water is a sensor to weak forces including electromagnetic fields of low intensity. Electromagnetic Biology and Medicine. 2005;**24**(3):449-461

[46] Teufel S. Adiabatic Perturbation Theory in Quantum Dynamics. Berlin, Germany: Springer; 2003

[47] Dykhne A. Quantum transitions in the adiabatic approximation. Soviet Physics—JETP. 1960;**11**:411

[48] Panagopoulos DJ, Johansson O, Carlo GL. Polarization: A key difference between man-made and natural electromagnetic fields, in regard to biological activity. Scientific Reports. 2015;**5**:14914

[49] Sowlati-Hashjin S, Matta CF. The chemical bond in external electric fields: Energies, geometries, and vibrational stark shifts of diatomic molecules. The Journal of Chemical Physics. 2013;**139**(14):144101

[50] Panagopoulos DJ, Messini N, Karabarbounis A, Philippetis AL, Margaritis LH. A mechanism for action of oscillating electric fields on cells. Biochemical and Biophysical Research Communications. 2000;**272**(3):634-640

[51] Panagopoulos DJ, Margaritis LH. Theoretical considerations for the biological effects of electromagnetic fields. In: Biological Effects of Electromagnetic Fields. Springer; 2003. pp. 5-33

[52] Leung K, Rempe SB. Ab initio molecular dynamics study of glycine intramolecular proton transfer in water. The Journal of Chemical Physics. 2005;**122**(18):184506

[53] Shamovsky I, Nudler E. New insights into the mechanism of heat shock response activation. Cellular and Molecular Life Sciences: CMLS. 2008;**65**(6):855-861

[54] Almlöf J. Hydrogen bond studies. 71. Ab initio calculation of the vibrational structure and equilibrium geometry in HF-2 and DF-2. Chemical Physics Letters. 1972;**17**(1):49-52

[55] Cassone G. Ab Initio Molecular Dynamics Simulations of H-Bonded Systems under an Electric Field. Université Pierre et Marie Curie-Paris VI; 2016

[56] Goryainov S. A model of phase transitions in double-well Morse potential: Application to hydrogen bond. Physica B: Condensed Matter. 2012;**407**(21):4233-4237

[57] Cerón-Carrasco JP, Jacquemin D. Electric field induced DNA damage: An open door for selective mutations. Chemical Communications. 2013;**49**(69):7578-7580

[58] Singh V, Fedeles BI, Essigmann JM. Role of tautomerism in RNA biochemistry. RNA. 2015;**21**(1):1-13

[59] Piilo J, Maniscalco S. Driven harmonic oscillator as a quantum simulator for open systems. Physical Review A. 2006;**74**(3):032303

[60] Chandra AK, Nguyen MT, Uchimaru T, Zeegers-Huyskens T. Protonation and deprotonation enthalpies of guanine and adenine and implications for the structure and energy of their complexes with water: Comparison with uracil, thymine, and cytosine. The Journal of Physical Chemistry A. 1999;**103**(44): 8853-8860

[61] Samdal S. The effect of large amplitude motion on the comparison of bond distances from ab initio calculations and experimentally determined bond distances, and on root-mean-square amplitudes of vibration, shrinkage, asymmetry constants, symmetry constraints, and inclusion of rotational constants using the electron diffraction method. Journal of Molecular Structure. 1994;**318**:133-141

[62] Rattan SI. Hormesis in aging. Ageing Research Reviews. 2008;**7**(1):63-78

[63] Rattan SI. Repeated mild heat shock delays ageing in cultured human skin fibroblasts. IUBMB Life. 1998;**45**(4):753-759

[64] Miura Y, Abe K, Urano S, Furuse T, Noda Y, Tatsumi K, et al. Adaptive response and the influence of ageing: Effects of low-dose irradiation on cell growth of cultured glial cells. International Journal of Radiation Biology. 2002;**78**(10):913-921

[65] Rattan SI. Mechanisms of hormesis through mild heat stress on human cells. Annals of the New York Academy of Sciences. 2004;**1019**(1):554-558

[66] Ben-Zvi A, Miller EA, Morimoto RI. Collapse of proteostasis represents an early molecular event in *Caenorhabditis elegans* aging. Proceedings of the National Academy of Sciences of the United States of America. 2009;**106**(35):14914-14919

[67] Udelsman R, Blake MJ, Stagg CA, Holbrook NJ. Endocrine control of stress-induced heat shock protein 70 expression in vivo. Surgery. 1994;**115**(5):611-616

[68] Shamovsky I, Ivannikov M, Kandel ES, Gershon D, Nudler E. RNA-mediated response to heat shock in mammalian cells. Nature. 2006;**440**(7083):556-560

[69] Perez FP, Moinuddin SS, ul ain Shamim Q, Joseph DJ, Morisaki J, Zhou X. Longevity pathways: HSF1 and FoxO pathways, a new therapeutic target to prevent age-related diseases. Current Aging Science. 2012;**5**(2):87-95

[70] Perez F, Millholland G, Peddinti SV, Thella AK, Rizkalla J, Salama P, et al. Electromagnetic and thermal simulations of human neurons for SAR applications. Journal of Biomedical Science and Engineering. 2016;**9**(9):437-444

[71] Durney CH, Massoudi H, Islanker MF. Radiofrequency Radiation Dosimetry Handbook. 4th ed. Brooks Air Force Base, TX: USAF School of Aerospace Medicine, Aerospace Medical Division (AFSC); 1986

[72] Huttunen P, Hanninen O, Myllyla R. FM-radio and TV tower signals can cause spontaneous hand movements near moving RF reflector. Pathophysiology. 2009;**16**(2-3):201-204

[73] Sefidbakht Y, Moosavi-Movahedi AA, Hosseinkhani S, Khodagholi F, Torkzadeh-Mahani M, Foolad F, et al. Effects of 940 MHz EMF on bioluminescence and oxidative response of stable luciferase producing HEK cells. Photochemical & Photobiological Sciences: Official Journal of the European Photochemistry Association and the European Society for Photobiology. 2014;**13**(7):1082-1092

[74] Rao DV, Watson K, Jones GL. Age-related attenuation in the expression of the major heat shock proteins in human peripheral lymphocytes. Mechanisms of Ageing and Development. 1999;**107**(1):105-118

[75] Hsu A-L, Murphy CT, Kenyon C. Regulation of aging and age-related disease by DAF-16 and heat-shock factor. Science. 2003;**300**(5622):1142-1145

[76] Cohen E, Bieschke J, Perciavalle RM, Kelly JW, Dillin A. Opposing activities protect against age-onset proteotoxicity. Science. 2006;**313**(5793):1604-1610

[77] Perez FP, Zhou X, Morisaki J, Ilie J, James T, Jurivich DA. Engineered repeated electromagnetic field shock therapy for cellular senescence and age-related diseases. Rejuvenation Research. 2008;**11**(6):1049-1057

[78] Shellock FG, Bierman H. The safety of MRI. Journal of the American Medical Association. 1989;**261**(23):3412

[79] Hill D. Human whole-body radiofrequency absorption studies using a TEM-cell exposure system. IEEE Transactions on Microwave Theory and Techniques. 1982;**30**(11):1847-1854

[80] Allen SJ. Measurements of power absorption by human phantoms immersed in radio-frequency fields. Annals of the New York Academy of Sciences. 1975;**247**(1):494-498

[81] Lin JC. A new IEEE standard for safety levels with respect to human exposure to radio-frequency radiation. IEEE Antennas and Propagation Magazine. 2006;**48**(1):157-159

[82] Perez FP, Ilie JI, Zhou X, Feinstein D, Jurivich DA. Pathomolecular effects of homocysteine on the aging process: A new theory of aging. Medical Hypotheses. 2007;**69**(1):149-160

[83] Park D, Jeong H, Lee MN, Koh A, Kwon O, Yang YR, et al. Resveratrol induces autophagy by directly inhibiting mTOR through ATP competition. Scientific Reports. 2016;**6**:21772

[84] Baur JA, Pearson KJ, Price NL, Jamieson HA, Lerin C, Kalra A, et al. Resveratrol improves health and survival of mice on a high-calorie diet. Nature. 2006;**444**(7117):337

[85] Small GW, Kepe V, Ercoli LM, Siddarth P, Bookheimer SY, Miller KJ, et al. PET of brain amyloid and tau in mild cognitive impairment. New England Journal of Medicine. 2006;**355**(25):2652-2663

[86] Perez FP, Bandeira JP, Morisaki JJ, Peddinti SVK, Salama P, Rizkalla J, et al. Antenna design and SAR analysis on human head phantom simulation for future clinical applications. Journal of Biomedical Science and Engineering. 2017;**10**(9):421-430

[87] Christ A, Chavannes N, Nikoloski N, Gerber HU, Pokovic K, Kuster N. A numerical and experimental comparison of human head phantoms for compliance testing of mobile telephone equipment. Bioelectromagnetics. 2005;**26**(2):125-137

[88] Suryadevara VK, Patil S, Rizkalla J, Helmy A, Salama P, Rizkalla M. Microwave/thermal analyses for human bone characterization. Journal of Biomedical Science and Engineering. 2016;**9**(02):101

[89] Rizkalla J, Jeffers M, Salama P, Rizkalla M. Electromagnetic simulation for diagnosing damage to femoral neck vasculature: A feasibility study. Journal of Orthopaedics. 2018;**15**(4):997-1003

[90] Borkar R, Rizkalla J, Kwon Y, Salama P, Rizkalla M. Electromagnetic simulation of non-invasive approach for the diagnosis of diabetic foot ulcers. Journal of Orthopaedics. 2018;**15**(2):514-521

Chapter 3

Study of Non-predictive Patterns of Non-Ionizing Radiation in the City of Salta in Argentine

Mario Marcelo Figueroa de la Cruz and Roberto Daniel Breslin

Abstract

Non-ionizing radiation (NIR) is a subject of continuous debate despite having been regulated internationally and at the level of organizations in all countries. This debate is focused on the level of population exposure to non-ionizing radiation density, since there is no certain evidence of the level of safety of the values adopted ranging from 0.2 mW/cm^2 to 0.2 uW/cm^2 according to the regulations of each state. The radiation precisions are made with models that evolve to take into account most of the factors that can attenuate the radiation emitted from an antenna from free space to models that take reflection and diffraction as attenuation factors. However, our work deepens in a phenomenon that is verified in measurement campaigns that is one of the values that do not fit with predictive models and that, on the contrary instead of showing attenuation, have higher values than expected. This work shows the results of observation campaigns of these points and their relationship with environmental conditions, which present diverse probabilities to explain their condition.

Keywords: non-ionizing radiation, population exposure, antennas, cellular telephony

1. Introduction

1.1 The characteristics of a normal radiation

A normal non-ionizing radiation (NIR) is that which is produced from a radiation mechanism based on electromagnetic propagation and its propagation components; that is, it follows a radiation mechanism where the electromagnetic wave encounters a discontinuity in its path waveguide, being forced to change the shape of longitudinal propagation within a closed environment to a spread radiation in an open environment.

The typical discontinuity that is used for the radiation to be efficient and allows controlling to some degree the propagation of an essentially isotropic radiation and transforming it into a directional radiation is the element called antenna.

An antenna, in essence, transforms an ideal isotropic radiation into a radiation that concentrates, in a certain direction, in the energy coming from a source of electromagnetic radiation. In this chapter, we analyze the characteristics of the normal radiations that generate predictive patterns based on the propagation characteristics and those of the antennas with respect to non-predictive patterns. To verify this alteration of the patterns, measurement points with discordant values were obtained in a specific campaign.

2. Elements of radiation

2.1 The antenna and its gain

The energy radiated by an antenna will be distributed uniformly in all directions and with divergent direction of the source, which in this case is the antenna, this case is also ideal since the radiation cannot be precise considering that the antenna does not it is more than a discontinuous prolongation of a transmission line and therefore has a feature of balanced line or balanced lines, this implies the existence of two elements or arms and therefore an antenna is basically a dipole.

This implies that, in essence, the radiation will have zero in the directions axial to the axis of that dipole.

In this way, if the dipole is placed with alignment to the Z axis, that is to say in a vertical position, the radiation null will be at ±90 of the dipole antenna itself.

In the case that the antenna is aligned with the axis y or x, the minimum radiation will be aligned with the corresponding axis.

However, the generality for broadcasting or cell phone transmission dipole antennas has an orientation on the Z axis in such a way that the main polarization is the so-called vertical polarization.

This means that the electric field vector has the same direction as the originating antenna, i.e., vertical (aligned with the Z axis), and that the magnetic field vector is of horizontal orientation since it will always be perpendicular to the vector of magnetic field enunciated in Maxwell's equations.

Radiation emitted by a dipole is conditioned by two factors that are not directly related to propagation or radiation and if the electrical characteristics of the circuit between a transmitting device and an antenna are:

- Adaptation efficiency
- Radiation efficiency

The adaptation efficiency depends on the adaptation of impedances between the transmission line and the antenna, where what is sought is that the impedance of the antenna constitutes the conjugate transpose of the impedance of the transmission line in which case the adaptation is perfect (**Figure 1**).

$$R_r + R_\Omega = R_g \qquad (1)$$

$$X_A = -X_g \qquad (2)$$

The radiation efficiency is specified by the relation between the totality of the energy that is delivered by the transmission line to the antenna and the energy actually radiated by it. The ideal situation is for the antenna to behave as an

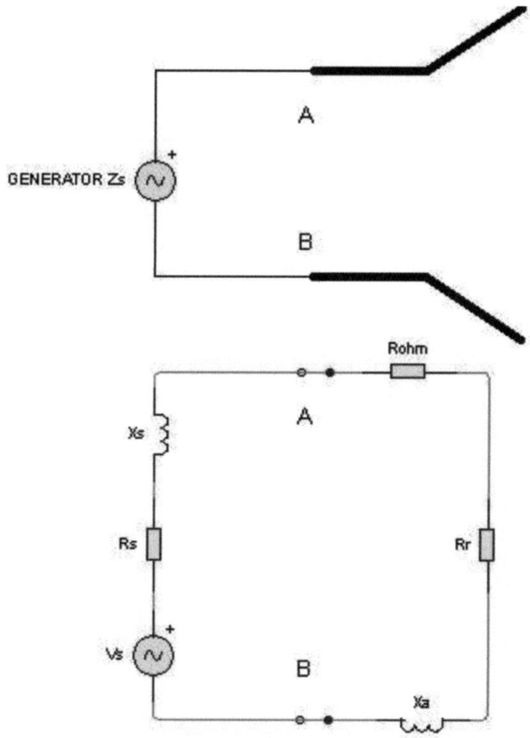

Figure 1.
Thevenin model of transmission of an antenna.

electrical element of resistance equal to zero, so that 100% of the energy delivered is radiated, this being an unrealizable situation; however, a part of the energy delivered is transformed into heat by Joule effect and is conditioned by the quality of the physical components of the antenna.

Therefore, the gain is conditioned by two efficiencies in such a way that the gain must be affected by the relationship between the impedance of the line and the antenna and the relationship between the ohmic and radiation resistances.

$$\eta_r = \frac{R_r}{R_r + R_\Omega} \quad (3)$$

$$Ca_T = \frac{P_A}{P_{AMAX}} = 1 - |\rho|^2 = 1 - \left|\frac{ROE - 1}{ROE + 1}\right|^2 \quad (4)$$

$$\rho = \frac{Z_L - Z_O}{Z_L + Z_O} \quad (5)$$

However, these considerations are not the only ones to be carried out for wireless communication and in particular for cellular telephony due to two factors:

- Limitation on profit
- Unequal coverage

2.2 The limitation of profit

The gain of an antenna is defined based on a factor called directivity.

The directivity is a factor that is given by the deformation of the radiation lobe that has a dipolar antenna and is measured as a relationship between a totally isotropic radiation and the radiation of a lobe that has an angular aperture bounded by a certain value less than that of isotropic radiation that has a 360-degree lobe.

The directivity can be defined as the lobe whose opening is limited by the angles that limit radiation with a drop of 3 decibels with respect to the maximum.

So the lobe will have a maximum of radiation in some direction and will have a smaller amount of radiation in directions with different angles from the maximum, that angle where it is verified that falls 3 decibels with respect to the maximum constitutes the limit of the radiation lobe or half power points. In short, the angle of opening will be twice the angle between the direction of the main lobe and the direction of the points of half power.

Obviously, the energy is not dispersed but is concentrated within the radiation lobe, i.e., the Poynting vector of the radiation lobe will have its maximum in the main direction of it and will decrease in different directions to that of the main lobe (**Figure 2**).

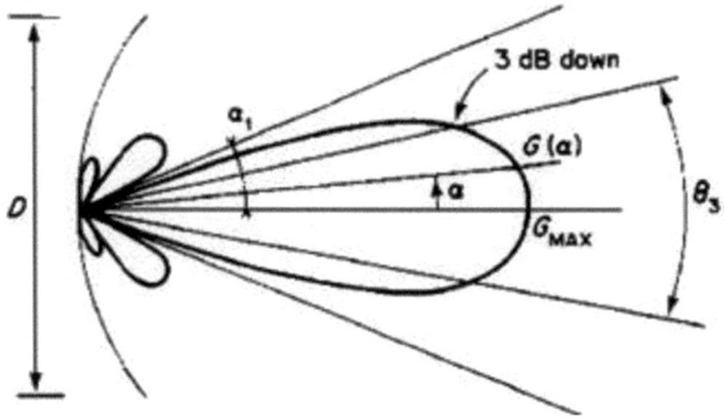

Figure 2.
Directivity of an antenna.

The gain of the antenna as previously expressed depends directly on the directivity so that the lower the opening of the lobe, the greater the verified directivity and, consequently, the greater the gain allocated to the antenna.

As a consequence, the Poynting vector has a much higher value in the main direction of the radiation lobe, where the greater part of the radiated energy is concentrated, so that to have greater radiated energy, a greater directivity is needed.

That is why what is sought in an antenna, in most cases, is the directivity, since having more energy in the center of the lobe, the attenuation will have a lesser effect and you can have a communication that arrives with good energy level at a greater distance.

The design of an antenna is based on the fundamental premise that the radiation lobe, both vertically and horizontally, is calculated for the electric field of the radiated wave.

However, the search in wireless communications is based on achieving the maximum use of radiated energy to achieve propagation distances as long as possible, which are why the gain of the antenna is the main factor in the efficiency of telecommunications.

2.3 Uneven coverage

Coverage is the geographical area in which the energy radiated by a radiator element such as an antenna has such a value that it can be exploited by an electronic receiving device.

That is, it is the area where the signal has a suitable value so that the receiving electronic equipment can adequately transform the electromagnetic signal into an electrical signal with an adequate signal-to-noise ratio in order to be intelligible information.

Depending on the type of coverage required, it will be the characteristic of the antenna that is used to radiate.

If what is required is an omnidirectional radiation, a dipole antenna is more than enough; however, it is precisely this antenna that has the lowest gain and, therefore, although it achieves an omnidirectional coverage, the attenuation it suffers is so strong that the coverage it has is very deficient.

That is why, for the omnidirectional coverage, antennas with a characteristic in the horizontal lobe of high aperture and in the vertical radiation lobe of high directivity are used.

A generic situation is to achieve the 360° coverage by antennas that have a horizontal lobe opening of 120° placing in this case three antennas or even placing 6 antennas with 60° openings.

As far as the coverage in distance is concerned, it depends on the vertical lobe opening and how it is required to achieve the widest possible coverage. It is required that the effects of the attenuation in the free space be compensated by the high gain or directivity that the antenna has.

For this, antenna panels with apertures smaller than 20° are used (**Figure 3**).

This coverage is usually referred to as the footprint of the antenna and the situation generates that in the areas closest to the antenna there is more radiation in the case that the center of the radiation lobe is aimed at the point closest to the antenna.

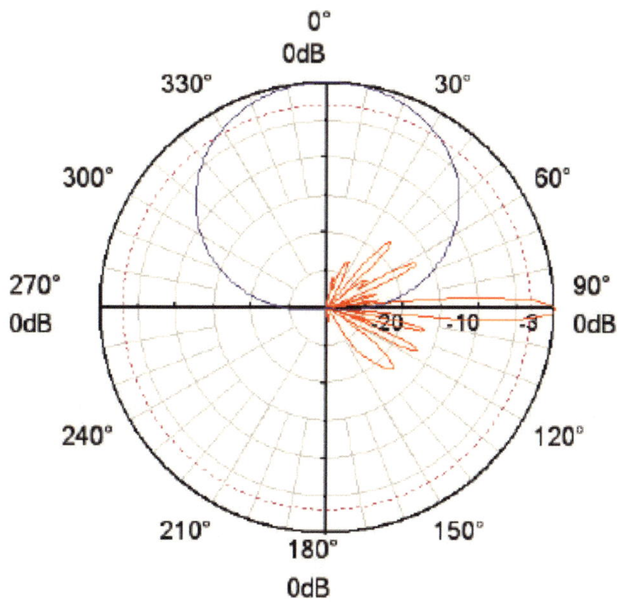

Figure 3.
Huawei Dual Band Panel Radiation Diagram ADU4518R3 [1].

What is required, for a good coverage, is that all users have reasonably the same level of signal. It is customary to make the center of the radiation lobe point toward the most extreme point, so that the signal level at the most extreme point coincides with the maximum of the radiation lobe. In the same way, those points that coincide with the minimum values of the radiation ovule, i.e., the 3 decibel points, are those that have less attenuation by distance.

In this way, the radiation lobe will tilt such that the level of radiation that will suffer the least attenuation by distance will fall at the point of coverage closest to the antenna and that the center of the radiation lobe having the highest power will suffer the greater attenuation in such a way that the coverage is equalized between both extremes and the users do not feel the decrease of the signal given by the attenuation in the free space (**Figure 4**).

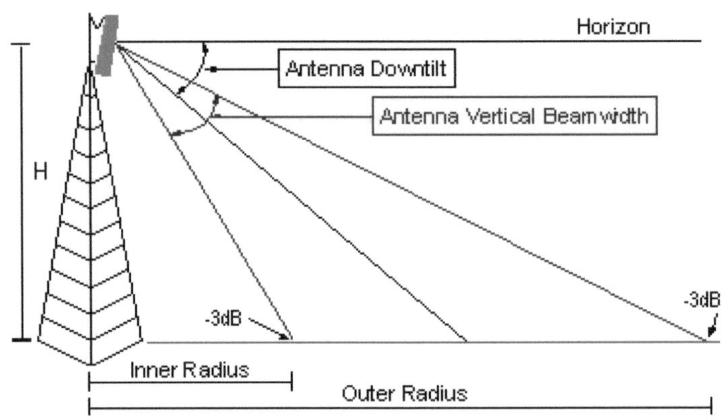

Figure 4.
Coverage method by downtilt [2].

It is evident that the height of the tower on which the antenna is located has a direct influence on the coverage area; however, there is a limitation that has to do with the attenuation in the free space so that the height of the tower has a practical limit that has to do with coverage and attenuation.

$$\frac{\text{Inner Radius}}{\text{Distance}} = \frac{H/\text{Tan}\left(A + \frac{BW}{2}\right)}{5280} \tag{6}$$

$$\frac{\text{Outer Radius}}{\text{Distance}} = \frac{H/\text{Tan}\left(A - \frac{BW}{2}\right)}{5280} \tag{7}$$

where A is the downtilt angle of the main beam, H is the height of the antenna relative to the ground, and BW is the angle of opening of the lobe for 3 dB of fall with respect to the main lobe.

These trigonometric equations give us the rule that the largest distance will depend on the angle of inclination that is the lowest, possibly and logically, and will also depend on the height of the tower as it has a direct proportionality.

It is very evident that if the value of the bandwidth of the beam width is 15° half of it will be 7° and the minimum distance of radiation from the base of the antenna would be corresponding to 7°. However, this would imply that it is directly pointing down. (A = 90°), which is not feasible or economically acceptable.

Considering that the Poynting vector analyzed practically how radiation density per unit area is 100 watts over square centimeters, a normal tread has a radiation density value of the order of 0.2 W/cm^2. This value is very important to take into account in the study of electromagnetic emission, derived from the radiation of a cell phone installation, characterizing the emission in the exclusive values of the frequency spectrum used by this application, for which the considered emission is in the values of 700 megahertz at 2.3 GHz

These values are directly related to the tread and with the radiated power from the radio bases and correspond to an average value of −79 decibels.

Mobile cellular equipment in turn will change its data reception system depending on the signal level in such a way that when receiving a signal lower than this value, it will automatically change to Enhanced Data Rates for GSM Evolution (EDGE) or General Packet Radio Service (GPRS) mode.

Therefore, the signal level will directly affect the speed of the internal modems of cellular equipment.

2.4 The antennas

Another aspect that has to be taken into account is what is normal radiation has to do with the emitting antennas.

The mobile telephone antennas are characterized by being bi-directional (emission or reception) of low power. In addition to producing RF radiation, they are mounted on poles, transmission towers, or the roofs of tall buildings, since they need to be at a certain height in order to have a wider coverage.

In a typical mobile telephone antenna, the radio emission is made toward the front and horizontally, in the form of a substantially flat beam, and covers a sector between 60 and 120°. Emissions are almost non-existent in the other directions (back, bottom, and top).

The characteristics of the antennas and the conditions in which they are usually installed make the emission levels in terms of radiation density very low in the place where they are located.

The flat panel antennas, as the name suggests, are a square- or rectangular-shaped panel. And they are configured in a patch-type format. Flat panel antennas are very directional since most of their radiated power is a single direction in either the horizontal or the vertical plane. In the elevation pattern (**Figure 5**) and in the azimuth pattern (**Figure 6**) [3], you can see the directivity of the flat panel antenna.

The flat panel antennas can be manufactured in different directivity values according to their construction. This can provide excellent directivity and considerable gain.

These panels, in fact, are an array of 4 or 5 antennas whose separation between each other within the panel and the different paths that run through the signal in the waveguides that feed them provide an additional tilt (called an electric tilt).

The angle of inclination of the main lobe is the sum of the mechanical TILT (conditioned by mobile supports) and the electric TILT given by the regulation of the paths (phase delays) of the power supply of each antenna of the array installed in a panel (**Figure 7**).

2.5 Calculation of normal radiation

Normal radiation follows a predictive pattern in which there are components that allow to know what is the power density value in the surface based on:

- The power with which the antenna is fed

- The attenuations of the connections

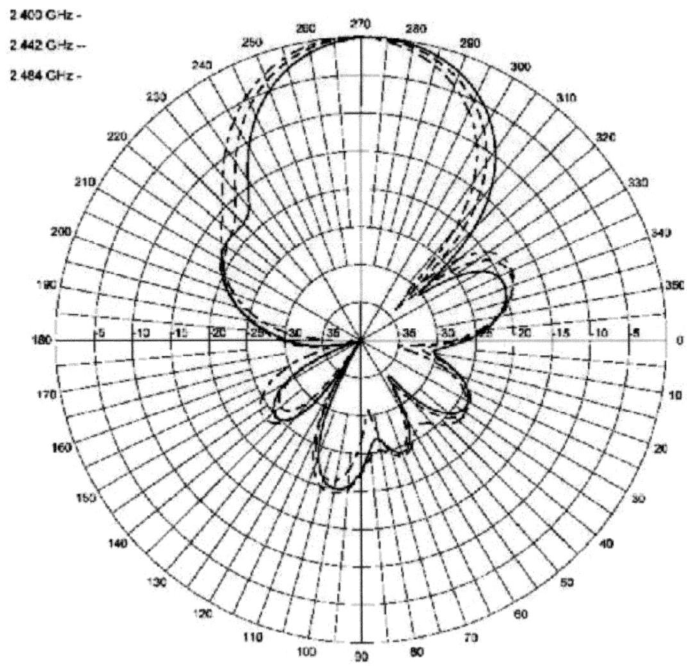

Figure 5.
Flat elevation pattern high gain panel [3].

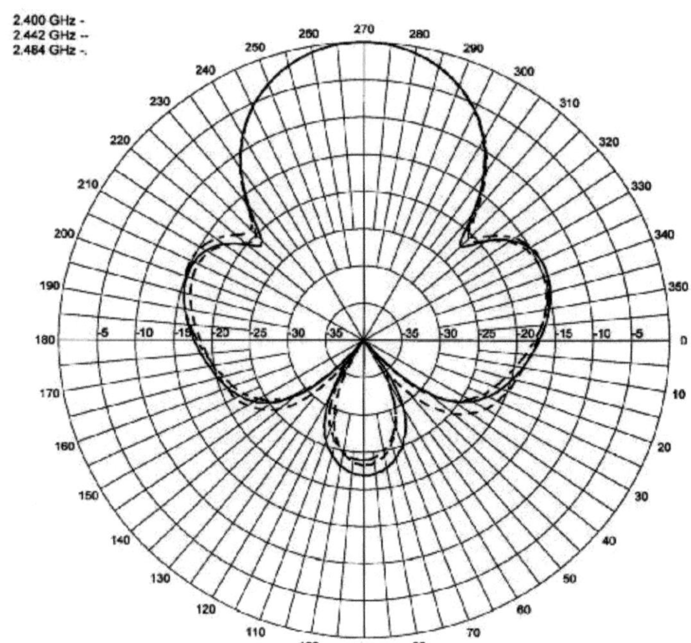

Figure 6.
Azimuth flat high gain panel pattern [3].

- The antenna gain affected by the coupling coefficients and radiation loss coefficient

- The attenuation in free space

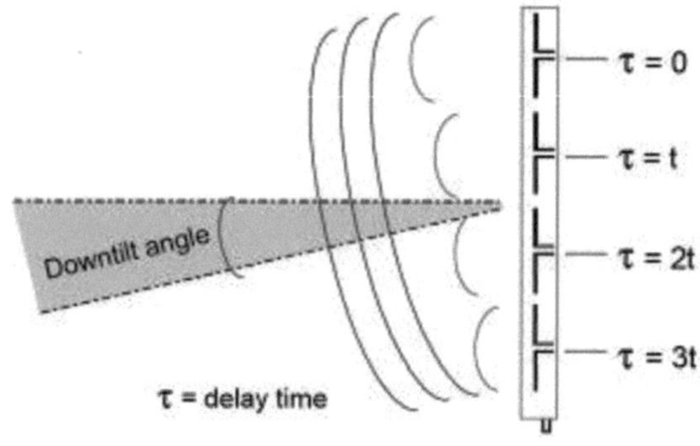

Figure 7.
Angle of an array of antennas caused by different phases [4].

The resulting value can still be added to a variety of attenuations produced by the surrounding environment that can be characterized as:

- High density urban
- Urban low density
- Suburban
- Rural

These attenuation parameters have been contemplated in various models such as the following:

2.5.1 Propagation model in free space

This model is used to predict the power of the signal when there is a clear line of sight between the transmitter and the receiver. Satellite communication systems and microwave links can be modeled as free space propagation. The free space model predicts that the received power decays as a function of the separation distance between the transmitter and receiver raised to some power. The power received in the free space by a receiving antenna, which is separated from the transmitting antenna by a distance d, is given by the equation.

$$\text{Pr (d)} = \text{PtGtGr } \lambda 2/(4\ \pi)2 \text{ d2 L} \tag{8}$$

Where Pt is the transmitted power, Pr is the received power that is a function of the Tx-Rx separation (transmitter-receiver), Gt is the transmit antenna gain, Gr is the gain of the receiving antenna, D is the separation distance of Tx-Rx given in meters, λ is the wavelength given in meters, and L depends on the obstacles, line of sight is = 1.

The equation shows that the power of the received signal is attenuated to form the square of the distance between the transmitter and the receiver, which implies that 20 dB /decade decays.

2.5.2 Okumura model

The Okumura model [5] is one of the most widely used for signal prediction in urban areas. This model is applicable for frequencies in the range of 150–1920 MHz, that is, it comprises the VHF and UHF band (however, it is typically extrapolated for frequencies above 3000 MHz entering the SHF band) and distances of 1 Km to 100 Km. It can be used for antenna heights of the base station in the range of 30–1000 m.

The model can be expressed as:

$$L50\ (dB) = LF + Amu\ (f, d) - G\ (hte) - G\ (hre) - GAREA \tag{9}$$

where L50 (dB) is the median attenuation per trajectory, LF is the free space attenuation, Amu (f, d) is the average relative attenuation (curves), G (htx) is the height gain of the Tx antenna, G (hrx) is the height gain of the Rx antenna, and GAREA is the gain due to the type of environment.

Okumura found that G (hte) varies at an index of 20 dB/decade and G (hre) varies at an index of 10 dB/decade for heights less than 3 m.

G (hte) = 20log (hte/200) for 30 m < hte < 1000 m

G (hre) = 10log (hre/3) for hre < 3 m

G (hre) = 20log (hre/3) for 3 m < the <10 m

It is one of the simplest and most suitable models for attenuation predictions.

2.5.3 Hata model (Okumura-Hata)

It is an empirical formulation of the propagation loss data provided by Okumura and is valid in the frequency range of VHF and UHF, from 150 to 1500 MHz. Although Hata [5] presented the losses within an urban area as a standard formula:

$$\begin{aligned} L50\ (urban)\ (dB) = 69.55 + 26.16\ \log\ fc - 13.82\ \log\ hte - a\ (hre) \\ + (44.9 - 6.55\ \log\ hte)\ \log\ d \end{aligned} \tag{10}$$

Taking into account that:

150 MHz < fc <1500 MHz

30 m < hte <200 m

1 m < hre <10 m

It should be considered that the definitions are the same as for the Okumura model, including:

- fc: carrier frequency [MHz].
- the height of the transmitting antenna in [m] in the range from 30 to 200 meters.
- hre: receiving antenna height in [m] in the range from 1 to 10 meters.
- a (hre): correction factor for the effective height of the mobile antenna that is function of the type of service area.
- d: distance between transmitter and receiver [km].

As can be seen, it involves a new variable that is the correction factor of the mobile antenna and is defined according to the size of the city:

For small- and medium-sized cities:

$$a\ (hre) = (1.1\ \log\ fc - 0.7)\ hre - (1.56\ \log\ fc - 0.8)\ dB \tag{11}$$

For large cities:

$$a\ (hre) = 8.29\ (\log\ 1.54hre)\ 2 - 1.1\ dB\ for\ fc < 300\ MHz \tag{12}$$

$$a\ (hre) = 3.2\ (\log\ 11.75hre)\ 2 - 4.97\ dB\ for\ fc > 300\ MHz \tag{13}$$

The following is the formula that can be used in a suburban environment:

$$L\ (dB) = L50\ (urban) - 2\ [\log\ (fc/28)]\ 2 - 5.4 \tag{14}$$

For rural areas:

$$L\ (dB) = L50\ (urban) - 4.78\ (\log\ fc)\ 2 + 18.33 \log\ fc - 40.94 \tag{15}$$

This model adapts very well for the design of large-scale systems, but not for PCS systems, which have cells of the order of 1 km radius. For this purpose, a numerical-empirical formulation of the graphical data provided by Okumura of attenuation for urban areas is made.

2.5.4 Model cost 231 (extension of the Hata model)

The European Cooperative for Scientific and Technical Research (EURO-COST) [5] developed the COST 231 model, in which it extends the Hata model up to the 2 GHz range covering the VHF and UHF bands. The model is expressed as:

$$\begin{aligned} L50\ (urban) = {} & 46.3 + 33.9\ \log\ fc - 13.82\ \log\ hte - a\ (hre) \\ & + (44.9 - 6.55\ \log\ hte)\ \log\ d + CM \end{aligned} \tag{16}$$

where CM is a correction factor to adapt the model by extending the frequency range for which the Hata model operates, CM is 0 dB for medium-sized cities and suburban areas, CM is 3 dB for metropolitan centers, and a (hre) corresponds to the equations presented in the previous topic (Hata Model).

One of the contributions of this model is to consider dispersion losses.

It is also defined in the following range:

f: 1500–2000 MHz

Th: 30–200 m

hre: 1–10 m

d: 1–20 km

2.5.5 Calculations based on normative resolution N° 3690/04

Although the above are mitigation calculations, the regulations applied in Argentina are those issued by CNC (National Communications Commission) under resolution number 3690 of 2004 [6].

In the aforementioned, reference is made to the fact that prior to any measurement, a predictive calculation based on attenuation in space must be made, which takes into account the antenna power to calculate the radiation density in relation to the distance.

$$S = \frac{PRA\ .1{,}64.2{,}56.F^2}{4.\pi.r^2} \tag{17}$$

where S is the power density in $\frac{W}{m^{2}}$, PRA is the antenna power in W, F is the attenuation at times of radiation at a certain angle of incidence in the vertical plane if it is unknown to adopt 1, 2.56 is an empirical reflection factor that takes into account the possibility that fields reflected in phase can be added to the direct incident field, and R is the distance from the antenna in meters.

3. Background on normal RNI measurements

Measurements of non-ionizing radiation are clearly developed not only with a number of works on the subject, only to exemplify the work of Azpurua et al. [7], the thesis work of Br. Jorge Juan Eduardo Ríos Solar [8], or the work that preceded it [9] and the relevant regulations in Argentina. In all these works and even regulations, measurements are made or measurements based on the theoretical radiation lobes are standardized of the installations taken as a reference for the emission of non-ionizing radiation.

3.1 Normal measurements made

This research developed a plan of field measurements, which focused on areas already studied in the project "Analysis of Measurements of Non-Ionizing Radiation in the City of Salta from the UCASAL" [9]. It was extended for 2 months in the areas studied and new areas.

The methodology emanating from RES N ° 3690/04 was used because the following stipulations were applied:

3.1.1 Selection criteria for measuring points

- The measurement must be made at accessible points by the public.
- The measurement points will be chosen according to the characteristics of the system.
- Irradiation and the wavelength of the emissions, where "applicable a the predictive calculations".
- For omnidirectional systems, at least 16 points should be selected, conveniently located on the ground, whose separation from the station is a function of the wavelength of the emitter.

Factors that influence the response of the instruments:
The following should be taken into account when making measurements:

- Variation of the impedance of antennas or probes in the vicinity of conductive surfaces
- Capacitive coupling between the probe and the field radiation source

The influence of these factors can be reduced by maintaining a separation greater than 20 cm or three times the size of the probe, whichever is greater, with respect to the source of re-irradiation field. That is why it is recommended that the antennas and/or probes are installed on tripods of non-conductive material.

3.1.2 Instruments used

In all cases, the instruments used were a TES-92 and TM-190 (TENMARS); previously, a contrast of measurements was made with an NARDA −550 that was taken as a calibration standard in a total of 25 measurements, the dispersion of results of 10% for TES-92 and 22% for TM-190, so the measurements presented correspond (unless specific indication) to TES-92.

The characteristics of the instrument are:

- Sensor type: electric field (E)
- Frequency range: 50–3.5 GHz
- Directional characteristic: isotropic, three-dimensional
- Measuring range (signal >50 MHz): 20 mV/m up to 108.0 V/m
- Error of use (@ 1 V/m and 50 MHz): ± 1.0 dB
- Frequency response (taking into account the number Factor CAL Factor: ±1.0 dB (50 MHz–1.9 GHz), ±2.4 dB (1.9–35 GHz)
- The noise deviation: Type. ±1.0 dB f > 50 MHz
- Load limit: 4.2 W/m^2 (40 V/m)

3.1.3 Normal measurement site Miguel Ortiz

The Miguel Ortiz site is an area centered on a monopost and located in the northern area of the city of Salta. It is characterized for being a suburban and quasi-rural location, since two of the panels partially cover areas corresponding to the field. It also covers an area of the Bolivian avenue that is an urban continuation of the highway of the same name. That is why in the path marked according to the regulations, there are rough areas of absolute clearance and line of sight (**Figures 8** and **9**).

Some deviations (attributed to motorway conditions) can be seen, although the trend of the curve follows a predictive pattern

3.1.4 Normal measurement site Castañares

This site is relevant because the first 150 ms of separation from the antenna are in free space due to the presence of a clear field; here, the installation is of a self-supporting tower 40 meters high (**Figures 10** and **11**).

3.1.4.1. Linear correction factor

A correction by footprint is made linearly from the point of −3 dB not taking into account the real lobe but an approximation that does not take secondary lobes into account.

$$K = 0.707 + (1 - 0.707)/(198 - 108).(\text{Distance} - 108) \qquad (18)$$

Figure 8.
Miguel Ortiz site measurement area.

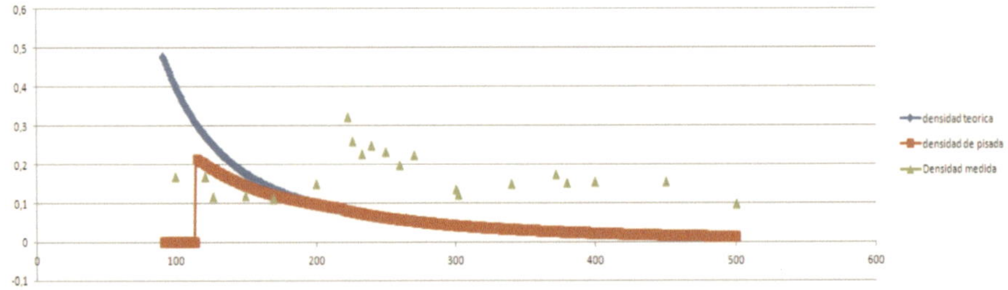

Figure 9.
Graph of radiation density measurements and predictive patterns. Miguel ortiz site.

Figure 10.
Castañares site measurement route.

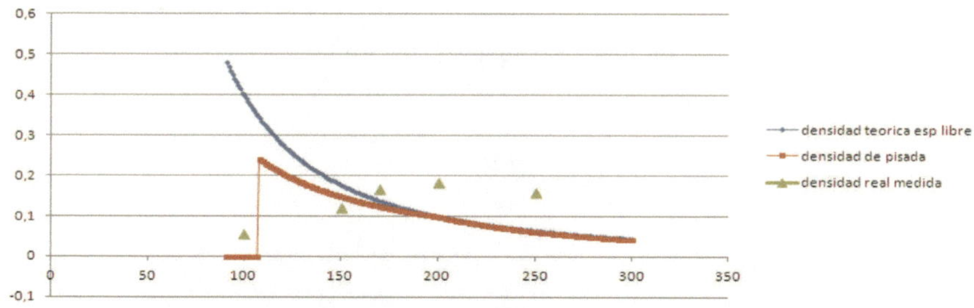

Figure 11.
Graph of radiation density measurements and predictive patterns. Castañares site.

3.1.5 Conclusions of normal measurements

In both cases, it can be observed that there is a clearance at least in the first section of the measurement and that it responds to the calculation trend taking into account the theoretical tread.

The uniformity in density values along the tread is also denoted, which is a desirable effect and compatible with an expected behavior for mobile telephony coverage.

4. Non-predictive patterns

During the measurement process referred to in the work "Analysis of Measurements of Non-Ionizing Radiation in the City of Salta from the UCASAL" [10], a series of random measurements were taken and also values were found that do not respond to a predictive pattern within urban trajectories and in trajectories in suburban environments but with sufficient clearance to be considered with little incidence of buildings; however, it was observed that there are phenomena that cannot be predicted with the stated models since they correspond to the attenuation in the free space and this is added in all cases of additional attenuations.

The cases that are listed show sites where, in the level of radiation density, not only does it not follow a predictive pattern, but also the level of radiation density increases.

4.1 History of discordant measurements

4.1.1 Aeroclub Salta site

The measurement in the Aeroclub Salta site constitutes the first measurement parameter discordant with the applied theory from the point of view of both the attenuation in the free space and the correction by tread (**Figures 12** and **13**).

The point located at 200 meters and the point located at 180 meters completely change the predictivity of the measurement; although the scope of measurement is free space in most of the path, the jarring factor is given in a point close to metal signage with a minimum height of 3 meters.

4.1.2 Guayacanes Site

This is a completely suburban site in the Tres Cerritos neighborhood of the City of Salta. It is a self-supporting tower 65 meters high and has a large number of panels as it is used by three telephone companies to provide service in the area.

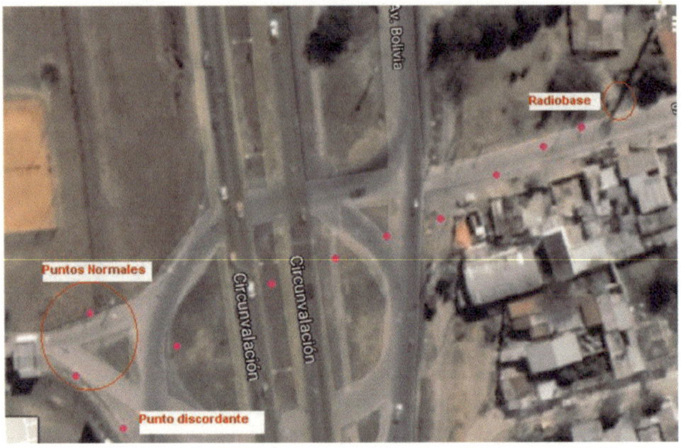

Figure 12.
Measurement trajectory and discordant points. Aeroclub Salta site.

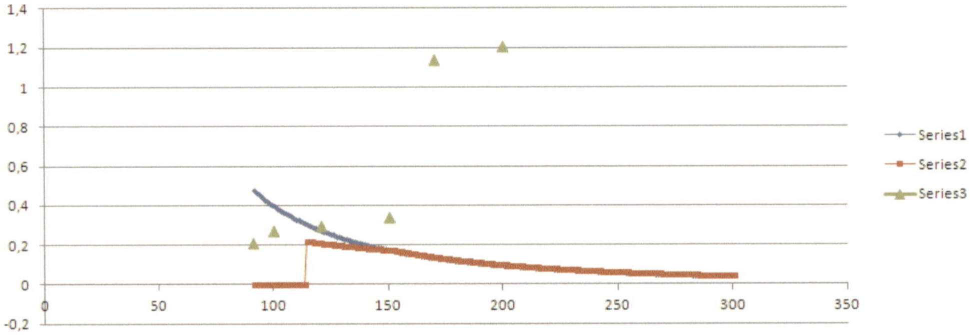

Figure 13.
Graph of radiation density measurements and predictive patterns. Aeroclub Salta site.

Measurements followed a path in a street (Las Acacias) that provides a line of sight of good quality and there is even a square with ample clearance (**Figures 14** and **15**).

The values measured in this area are much higher than the trend, and additionally, they are punctual since, as the advance of the planned measurement path continues, they return to their normal value.

It can be seen that in a high clearance area there are highly discordant points

Figure 14.
Measurement trajectory and discordant points. Guayacanes site.

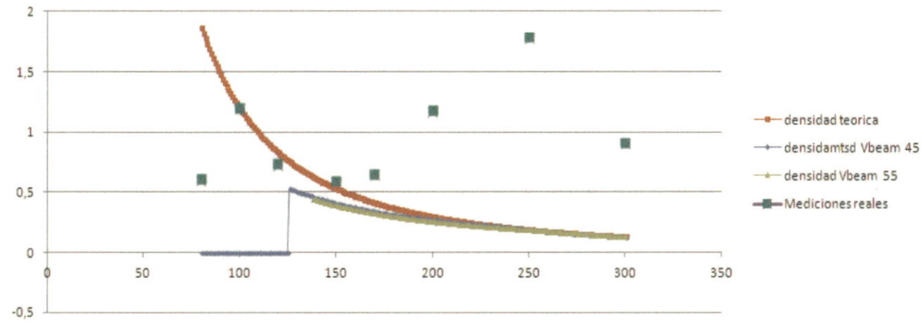

Figure 15.
Graph of radiation density measurements and predictive patterns. Guayacanes site.

4.1.3 Considerations regarding low-frequency radiation

The aerial lines of low- and medium-voltage lines are in some cases considered as causing a probability of increase in the radiation density by radiofrequency, which is why it is worth considering what the applicable concept of immission is and what the ranges of measurement of the immission sensors used by the instruments are.

4.1.3.1 Definition of Immission

As can be inferred from [11], "this term refers (in law) to an attack, aggression or attack of environmental type or also to a concentration, agglutination or conglomeration of pollution or the transfer of pollutants in a place or site and at a specific moment, more in common in the air and in general to an electromagnetic energy or radioactive particles."

The analysis of the definition is very useful because the agglutination does not imply a sum necessarily (it is an essentially biological concept) [12] and the possibility of a concentration.

In the case of electromagnetic immission, it must be considered that the sensors used for measurement are made in the three electric field vectors (x and z) and the sum is a vector sum of these fields.

However, the sensed fields start from frequencies from 100 Khz. If we consider the harmonic spectrum of an electrical network, this would correspond to a harmonic of order 2000, considering 50 Hz as fundamental frequency, so it cannot be considered to the radiation of electric lines as a member of the emission that can be measured by RNI sensors.

Such is the case that there are no sensors with such a large bandwidth to detect RF radiation with values below 100 KHz [13], and in fact the detectors are based on Schottky diodes with bounded bandwidths, hence the need to change the probe for certain high-frequency measurements.

5. Measurements of points that do not respond to predictive patterns

5.1 Work methodology

In the research project "Analysis of non-predictive patterns of non-ionizing radiation in a sector of the northern area of the City of Salta" approved by the research council of the UCASAL a plan of measurements inside and outside lobe trajectories is planned of radiation, a work methodology and a measurement protocol were implemented in accordance with the objective of identifying and characterizing points where the radiation value exceeds the normal value of the area. Off-trajectory measurements were also made from radiation lobes centered on radio bases with the sole purpose of verifying the possible existence of other points of discordance, not with a radiation pattern but with their close environment.

Thus, random routes were generated with a permanent measurement methodology with configurable alarm instrument, and it was particularized in density levels higher than 1 uW/cm^2.

For the field measurements, both the TES-92 meter (which exclusively measures the emission in a range of 300 KHz–2.5 GHz) and the TM-190 meter that measures a range of 50 MHz–3.5 GHz of RNI were used. And additionally, it measures electromagnetic and electric field radiation separately in the frequency of 50 to 60 exclusively. The objective of this double measurement was to contrast the presence of abnormal values of electric fields in conditions of electromagnetic fields relevant to this study.

5.2 Background of "hot" points

In informal measurements prior to the launch of the measurement campaign, 4 points were found where the level greatly exceeds the level of 1 uW/cm^2, reaching values of 2.75 uW/cm^2. That measurement campaign was launched with the aim of finding other points and analyzing the environmental conditions in order to establish a prediction pattern.

5.3 Work hypothesis

5.3.1 Metallic surfaces

As a consequence of the aforementioned antecedents, the possibility arises that there are specific locations, or small areas where phenomena of increase in the level of radiation density are manifested that are not a consequence of the emission produced by antenna radiation, but of other factors which are related to the environment of the point in question.

There is sufficient evidence of the increase in light radiation at infrared frequencies and visible by the effects of the surfaces on which they affect, such as the case of Raman scattering. [14] When light interacts with matter, it can disperse inelastically from vibrational quantum states. During this process, the photons can lose energy, or gain it from vibratory excitations, and it can also produce a concomitant change in the scattered frequency. The phenomenon, called the Raman effect, was discovered experimentally in 1928 by C. V. Raman and K. S. Krishnan in India and, independently, by Leonid Mandelstam and Grigory Landsberg in the former Soviet Union.

Oldenburg SJ, Hale GD, Jackson JB, and Halas NJ postulate in their work Light scattering from dipole and quadrupole nanoshell antennas [15] that metal nanoshells are nanoscale optical components that allow the controllable redirection of electromagnetic radiation through a careful engineering of its multilayer structures. By varying the size of the core and the thickness of the shell of these nanoparticles, nanoscale "antennas" are constructed that can be selectively driven in a dipolar or quadrupole oscillation pattern. With spaced transverse sections many times larger than their physical cross-section, these antennas are efficiently coupled to the incident electromagnetic wave. These structures can focus, redirect, or divide incident light

In this sense, Martin Moskovits in his work Surface-Enhanced Spectroscopy [16] reveals that molecules adsorbed on specially prepared silver surfaces produce a Raman spectrum that is sometimes a million times more intense than expected. This effect was called improved surface Raman scattering (SERS). Since then, the effect has been demonstrated with many molecules and with several metals, including Cu, Ag, Au, Li, Na, K, In, Pt, and Rh.

Another factor that could be the cause of non-predictive patterns is stated in the work of Francisco J. Rodríguez-Fortuño, Giuseppe Marino, Pavel Ginzburg, Daniel O'Connor, Alejandro Martínez, and Gregory A. W, Near-Field Interference for the Unidirectional Excitation of Electromagnetic Guided Modes [17], where they postulate that wave interference is a fundamental manifestation of the superposition principle with numerous applications. Although in conventional optics, the interference between waves that experience different phase advances during propagation, the vector structure of the near field of an emitter is essential to control its radiation, since it interferes with the interaction with a mediating object. Then, the near-field interference of a circularly polarized dipole results in the unidirectional excitation of the electromagnetic modes guided in the near field, without a preferred far-field radiation direction.

With these studies, it can be postulated that in the vicinity of certain surfaces there may be abnormal concentrations of radiation with respect to a prediction based on the propagation models from a source based on an antenna and that there is a dependence of the characteristics of this abnormality with the polarization that occurs in the near field of the source, so not all antennas could produce constructive interferences.

Under this hypothesis is that a campaign of measurements focused on the probability of finding hot spots of radiation density in certain environmental conditions, in particular, metallic reflective surfaces or with reflective paints, is required.

It is a possibility that these points are the consequence of antennas or panels that have circular polarization characteristics, that is, discarding the linear polarization dipoles.

5.3.2 Electric distribution transformers

Although the emission should not be related to power lines, it is no less true that the process of transforming voltage with high power levels could generate harmonics that contribute to the emission, whose value could be a significant analysis of the presence of transformers in the vicinity of points of high radiation density.

5.3.3 Buildings of 2 or 3 floors that could produce reflections

Resolution ITU No P526-11 is a methodology that includes a series of predictive patterns taking into account a variety of effects of buildings on the radiation pattern by diffraction [18] and most other propagation models also take into account reflection effects. It is necessary to identify the presence of buildings with a differential height with respect to the normal level of buildings; in this sense, it will take into account only the buildings very close to the points of differential radiation of electromagnetic density.

For these purposes, a non-predictive non-ionizing radiation measurements protocol was designed.

6. Protocol of measurement of RNI under non-predictive patterns

In order to systematize the analysis of the environmental conditions that could eventually generate points with differential RNI levels, a working protocol was established based on the hypothesis that certain elements or constructive characteristics could be generating the points with radiation differentials with respect to normal radiation in the area.

6.1 Measurement protocol of non-predictive patterns

1. A known radiation source of omnidirectional radiation must be determined in the UHF or higher bands whose radiation model is omnidirectional. To this end, the application Cell Network Info lite or similar should be used in an Android phone to locate the antenna on which the route will be based.

2. An analysis grid will be drawn identifying the route to be carried out. For this purpose, radial routes to the location of the tower should be prioritized.

3. A walking tour will be started by holding the TES 92 meter or similar in a hand at an approximate height of 1.60 meters above ground level, your own cell phone or any other device that can emit radiation (e.g., Bluetooth) must be deactivated.

4. The field analyst must observe the radiation indicator At all times looking for radiation patterns that exceed the normal nominal values of the area by 100% or more.

5. Upon the detection of a value such as stated above, proceed as follows:

   a. The field analyst will stop or start a scan in the detailed area looking for the point where the maximum radiation is detected.

   b. At that point, he will proceed to do the following:

      i. Record the exact coordinates of the site.

      ii. Photograph the meter at that point.

      iii. Take the meter to a nearby point where the radiation level is significantly lower.

      iv. Record the coordinates of this point.

      v. Photograph the meter in the foreground and in the background the place where the significant level was verified.

      vi. Return to the point where the significance was verified.

   c. He will make an observation of the physical/constructive characteristics of the place, emphasizing the following aspects:

      i. Constructions or parts of metal buildings both solid and grilled.

      ii. Metal signage nearby.

      iii. Surfaces that can be identified as reflective.

      iv. Metal gates.

v. Vehicles parked in the vicinity (in this case, check again when the vehicle is not).

vi. Painted surfaces with some type of metallic or shiny paint.

vii. Presence of obvious radiation sources (such as antennas, transformers, welding machines, electric motors in operation, generators, etc.).

viii. Any other aspect of the environment that is relevant.

d. He will photograph the surrounding environment and place particular focus on the constructions or elements listed.

6. Continue the journey under the rules stated.

7. Results of measurements under protocol of measurement of RNI under non-predictive patterns

Table 1 shows the values found taking into account the value of the point found and the value of its environment, obtaining the differential value. The table is ordered in descending order with respect to the differential.

7.1 Analysis of sites with non-predictive values of radiation level

From the analysis of **Table 1**, the following conclusions can be obtained:

- The radius of a point of high differential radiation is variable and does not depend on the center's RNI level.

- The radiation differential between the maximum and the normal level is not directly related to the maximum level of radiation.

- The observed average radius is 1.81 mts.

7.2 Analysis of the relationship between points of high level of radiation and distance to the most likely site of emission

Table 2 is shown in order to determine the possible relationship between the proximity to the emission source and the points with the highest differential value of radiation density with respect to the normal values of the site.

7.2.1 Conclusions of analysis of the relationship between points of high level of radiation and distance to the most likely site of emission

From **Table 2** and **Figure 16**, it can be seen that there is no relationship between the increase in the differential of radiation and the distance to the most probable source of radiation since, for example, for a distance to the site of most likely emission of 100 meters, radiation differential values as different as 16.95 uW/cm^2 $\frac{uW}{cm^2}$ are appreciated as 0.81 uW/cm^2.

ID site	Coordinates	Dif real-normal	Level l $\frac{uW}{cm^2}$	Normal level $\frac{uW}{cm^2}$	Radius (mts)
9-Rioja 880	−24.76664, −65.39294	16.95	20.15	3.2	0.5 a 1
14-Avda. Independencia 1286	−24.80827, −65.39923	16.14	17.64	1.5	4
8-Rioja 862	−24.77988, −65.40314	7.71	8.51	0.8	0.4 a 0.9
12-Avda. Paraguay y J Castellanos	−24.80575, −65.4197	4.134	5.058	0.924	2
15-Avda. Independencia 1290	−24.80828, −65.39889	4.05	5.25	1.2	3
7-Rioja 842	−24.7813, −65.40111	2.859	3.599	0.74	0.9 a 1.5
4-Florida y San Luis	−24.77664, −65.40431	2.772	4.011	1.239	0.4 a 1
5-Florida 602	−24.7796, −65.40132	2.609	3.223	0.614	0.3 a 07
Avda. Uruguay 735	−24.7813, −65.40111	2.026	3	0.537	4
20-Mendoza “Lago del Parque”	−24.79541, −65.4028	1.602	3.152	1.55	2
16-Avda. Independencia 1326	−24.80825, −65.39829	1.26	2.06	0.8	2
Avda. del Bicentenario 800	−24.77988, −65.40314	1.19	2	0.33	5
25-Sarmiento y 12 de Octubre	−24.77378, −65.41483	1.161	1.791	0.63	1
18-Santa Fe y Mendoza	−24.79534, −65.4045	1.138	1.738	0.6	2
2-Santa Esq. Rioja	−24.76743, −65.39803	1.067	2.533	1.466	0.5 a 1
19-Mendoza 2	−24.79551, −65.40302	0.935	1.735	0.8	3
Avda. Uruguay 895	−24.7796, −65.40132	0.9	1	0.253	1
23-Aniceto Latorre y A. Güemes	−24.77307, −65.41625	0.883	1.57	0.687	1
22-San Martin Y Lavalle	−24.79374, −65.40344	0.85	1.45	0.6	3
26-Sarmiento y 12 de Octubre	−24.77416, −65.41455	0.811	1.261	0.45	2
Vicente López 1000	−24.77664, −65.40431	0.81	1	0.22	4
24-Aniceto Latorre y A. Güemes	−24.77309, −65.41596	0.719	1.029	0.31	1
11-Lamadrid y Lola mora	−24.80426, −65.42133	0.7	1.3	0.6	1
Los Jazmines 840	−24.76403, −65.3905	0.667	0.777	0.11	3

ID site	Coordinates	Dif real-normal	Level l $\frac{uW}{cm^2}$	Normal level $\frac{uW}{cm^2}$	Radius (mts)
Avda. Uruguay 751	−24.78087, −65.4012	0.663	0.955	0.292	2
Vicente López 1075	−24.77694, −65.40427	0.621	0.934	0.313	2
6-Rioja 710	−24.78087, −65.4012	0.601	1.141	0.54	0.5 a 1
Los Jazmines y Los Mandarinos	−24.76664, −65.39294	0.587	0.6	0.013	10
21-Mendoza "Paseo"	−24.79544, −65.40224	0.511	1.523	1.012	2
1-Santa Fe 698	−24.76403, −65.3905	0.421	1.071	0.65	1 a 1.5
13-Avda. Independencia y calle Pucha	−24.80825, −65.40466	0.4	0.7	0.3	1
17-Avda. Independencia 1360	−24.80827, −65.39797	0.3	0.5	0.2	2
Av. Reyes Católicos 1617	−24.76743, −65.39803	0.294	0.535	0.241	1
3-Rioja Esq. Catamarca	−24.77694, −65.40427	0.21	0.89	0.68	0.3 a 0.7
Los Ombúes 95	−24.76449, −65.39652	0.129	0.226	0.097	3
10-Jujuy 804	−24.76449, −65.39652	0.06	0.61	0.55	0.4 a 0.8

Table 1.
Results of samples under non-predictive pattern measurement protocol.

7.3 Analysis of the incidence of the proximity of transformers of electrical distribution with respect to the differential of radiation density

7.3.1 Conclusions of the analysis of the incidence of the proximity of transformers of electrical distribution with respect to the differential of radiation density

From **Table 3** and **Figure 17**, it can be concluded that the presence of electrical distribution transformers does not constitute a conditioning factor for the point increase in the level of differential radiation density. However, the lack of influence in the specific increase cannot be guaranteed since they have an incidence of 27% in the total of events.

7.4 Analysis of the incidence of the proximity of billboards, gates, or metal fences with respect to the differential of radiation density

7.4.1 Analysis of the incidence of the proximity of signage, gates, or metal fences with respect to the differential of radiation density

According to working hypothesis 4.3, the analysis of metal surfaces near points of high differential level of radiation density is of particular interest.

Real-normal dif of S in $\frac{uW}{cm^2}$ x 100	Theoretical radiation source (mts)
1695	100
1614	15
771	80
413.4	200
405	20
285.9	70
277.2	200
260.9	100
202.6	38
160.2	250
126	30
119	86
116.1	245
113.8	200
106.7	70
93.5	220
90	194
88.3	370
85	400
81.1	200
81	94
71.9	360
70	3
66.7	400
66.3	63
62.1	90
60.1	150
58.7	400
51.1	250
42.1	50
40	300
30	50
29.4	300
21	50
12.9	420
6	300

Table 2.
Differential radiation density values as a function of distance to probable source.

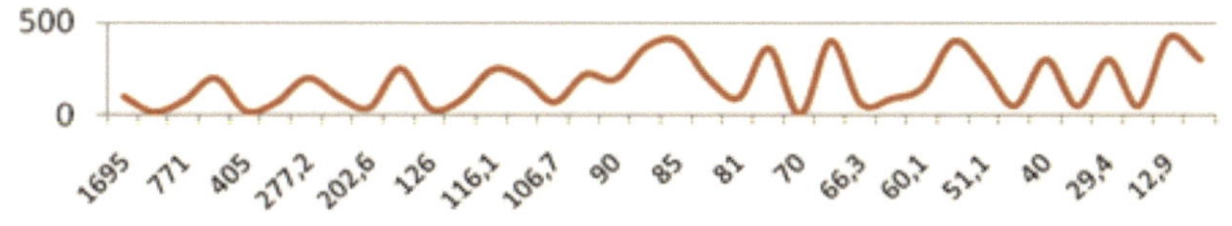

Figure 16.
Differential radiation density as a function of distance to probable source.

From **Table 4**, the following can be concluded:

- Some surface 89%
- Metal posters 27%
- Metal gates 57%
- Fences 68%

The probability increases by 30% if there is more than one metal surface in the vicinity (**Figure 18**).

Having some type of metallic surface, there is an 89% chance of finding a point with radiation density differential (**Figure 19**).

7.5 Analysis of the relationship between nearby buildings and the radiation level differential

7.5.1 Analysis of the relationship between nearby buildings and the radiation level differential

From **Table 5**, the following can be concluded:

- Some building 84%
- Buildings of 2 floors 73%
- Buildings of 3 floors or more 22%

In 84% of the cases of differential increase in radiation density, there is a building with 2 or more floors nearby, although the presence of these two types of buildings only increases the possibility by 0.06%.

7.6 Other factors analyzed

The following additional factors have been taken into account as a probable factor of point increase:

- Influence of vegetation and trees
- Pavement
- Metallic ceilings
- Electrical wiring
- High density of parked vehicles

Real-normal differential $\frac{uW}{cm^2}$	Proximity of transformers (mts)
16.95	1
16.14	0
7.71	1
4.134	0
4.05	0
2.859	1
2.772	0
2.609	0
2.026	0
1.602	0
1.26	1
1.19	1
1.161	0
1.138	0
1.067	0
0.935	0
0.9	0
0.883	0
0.85	0
0.811	0
0.81	0
0.719	0
0.7	1
0.667	0
0.663	0
0.621	0
0.601	0
0.587	0
0.511	0
0.421	0
0.4	0
0.3	1
0.294	0
0.21	1
0.129	1
0.06	1

Table 3.
Presence of distribution transformers in relation to differential level of radiation density.

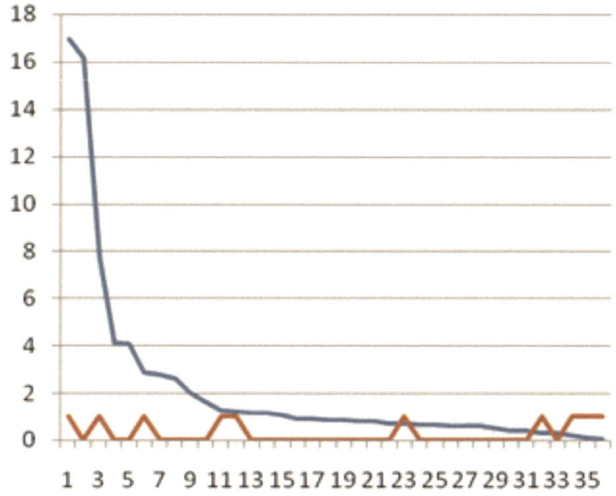

Figure 17.
Cases of presence of distribution transformers in relation to differential level of radiation density (y axis).

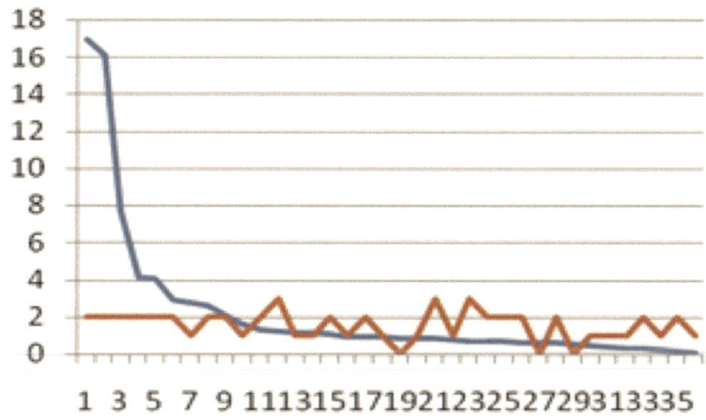

Figure 18.
Incidence of the number of cases of metallic surfaces with respect to the differential of radiation density.

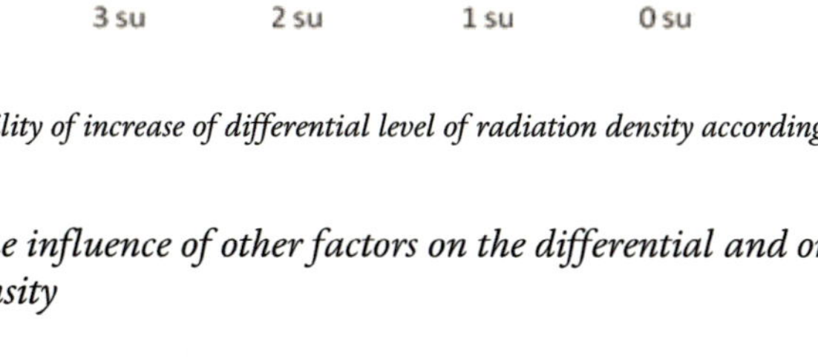

Figure 19.
Distribution of probability of increase of differential level of radiation density according to the amount of metal surfaces nearby.

7.6.1 Analysis of the influence of other factors on the differential and on the point value of radiation density

In this analysis, it is possible to deepen the point radiation according to its value. From the analysis of **Table 6**, the following can be concluded:

Differential real-normal $\frac{uW}{cm^2}$	**High metal signage cases**	**Metal gates cases**	**Metal fences/wiring cases**	**Number of cases**
16.95	0	1	1	2
16.14	0	1	1	2
7.71	0	1	1	2
4.134	1	0	1	2
4.05	0	1	1	2
2.859	0	1	1	2
2.772	0	0	1	1
2.609	0	1	1	2
2.026	0	1	1	2
1.602	0	0	1	1
1.26	1	1	0	2
1.19	1	1	1	3
1.161	0	0	1	1
1.138	1	0	0	1
1.067	1	1	0	2
0.935	0	0	1	1
0.9	1	1	0	2
0.883	0	0	1	1
0.85	0	0	0	0
0.811	0	0	1	1
0.81	1	1	1	3
0.719	0	0	1	1
0.7	1	1	1	3
0.667	0	1	1	2
0.663	0	1	1	2
0.621	0	1	1	2
0.601	0	0	0	0
0.587	0	1	1	2
0.511	0	0	0	0
0.421	0	1	0	1
0.4	0	0	1	1
0.3	0	1	0	1
0.294	0	1	1	2
0.21	1	0	0	1
0.129	0	1	1	2
0.06	1	0	0	1

Table 4.
Closeness of billboard, gates, or metal fences with respect to the differential of radiation density.

Real-normal differential	2-floor buildings	Buildings of 3 or more floors	Amount	Amount > 0
16.95	1	0	1	1
16.14	1	0	1	1
7.71	1	0	1	1
4.134	1	1	2	1
4.05	1	0	1	1
2.859	1	0	1	1
2.772	1	1	2	1
2.609	1	0	1	1
2.026	1	0	1	1
1.602	1	0	1	1
1.26	1	0	1	1
1.19	0	1	1	1
1.161	0	0	0	0
1.138	0	0	0	0
1.067	1	0	1	1
0.935	1	0	1	1
0.9	1	0	1	1
0.883	0	1	1	1
0.85	0	0	0	0
0.811	0	0	0	0
0.81	1	1	2	1
0.719	0	1	1	1
0.7	1	0	1	1
0.667	1	0	1	1
0.663	1	0	1	1
0.621	1	0	1	1
0.601	1	0	1	1
0.587	1	0	1	1
0.511	0	0	0	0
0.421	1	0	1	1
0.4	1	0	1	1
0.3	1	0	1	1
0.294	0	1	1	1
0.21	1	0	1	1
0.129	1	0	1	1
0.06	1	1	2	1
Total	27	8		31

Table 5.
Relationship between nearby buildings and the radiation level differential.

Real-normal differential	High-density electrical wiring	High-density parked vehicles	Site under plants	Metal ceilings nearby
16.95	1	1	1	0
16.14	0	0	1	1
7.71	1	1	1	0
4.134	0	1	0	0
4.05	0	0	1	1
2.859	1	1	1	0
2.772	0	0	0	0
2.609	0	1	0	1
2.026	0	0	0	0
1.602	0	0	0	0
1.26	1	0	0	0
1.19	1	1	0	1
1.161	0	0	0	0
1.138	0	0	0	0
1.067	0	1	0	0
0.935	0	0	0	0
0.9	0	1	0	0
0.883	0	1	0	1
0.85	0	0	0	1
0.811	0	0	0	0
0.81	0	1	1	0
0.719	0	1	0	1
0.7	1	0	0	1
0.667	0	0	0	0
0.663	0	1	0	1
0.621	1	1	0	0
0.601	0	0	0	0
0.587	0	0	0	0
0.511	0	0	0	1
0.421	0	1	0	0
0.4	0	0	1	0
0.3	1	0	0	0
0.294	1	0	0	0
0.21	1	0	1	0
0.129	1	0	0	0
0.06	1	1	0	0

Table 6.
Influence of other factors on the increase in the difference in radiation density.

High levels:

- 63% cars
- 63% plants
- 38% roofs

Low levels:

- 34% cars
- 10% plants
- 24% ceilings

The presence of cars parked in the vicinity, plants or trees on the measurement area, and to lesser extent metal roofs have a strong influence on the point value, such that in the values above the average of 2.12 uW/cm^2, the presence of these factors is verified. On the other hand, in low radiation values, there are no such conditions. Therefore, it can be concluded that vehicles and some types of plants contribute to the possibility of an increase in the level of radiation density.

It is also observed that at the highest levels there is the presence of two of these conditions.

8. Conclusions

There is no doubt that electromagnetic radiation has a number of variables whose study is constantly evolving, at the same time as technology evolves. New radiation patterns are incorporated to make the calculation more accurate and the behavior of electromagnetic waves more predictive.

In this work, it is demonstrated by field exploration in urban and suburban environments that there are points of high concentration of radiation, that these points have locations not related to the distance to the source, which are only points and not extensive areas with smaller radii than 4 meters, and that you can evaluate possibilities of existence of these points based on the surrounding environment, taking into account metallic surfaces at level and above ground level, constructions of more than one plant nearby, and to a lesser extent other factors, but in some cases, they are definitive in the location of what can be called hot spots of differential density of non-ionizing radiation.

Acknowledgements

The authors wish to express their gratitude to the Research Council of the Catholic University of Salta (UCASAL) for their support in the development of the research.

Conflict of interest

The authors declare that there is no potential conflict of interest related to the article.

Author details

Mario Marcelo Figueroa de la Cruz[1,2]* and Roberto Daniel Breslin[1]

1 Catholic University of Salta (UCASAL), Salta, Argentina

2 National Technological University, Tucuman, Argentina

*Address all correspondence to: mfiguero@gmail.com

References

[1] ADU4518R3. DXX-790-960/1710-2180-65/65-16.5i/18.5i-M/M-R. EasyRET Dual-Band Antenna with 2 Integrated RCUs. Electrical. Frequency range (MHz). Available from: https://www.huawei.com/ucmf/groups/public/documents/attachments/hw_316934.pdf [Accessed: 05-01-2019]

[2] Wireless Calculators. Available from: http://www.terabeam.com/support/calculations/index.php [Accessed: 05-01-2019]

[3] WNI Mexico Types of Antennas and Operation. Available from: https://www.wni.mx/index.php?option=com_content&view=article&id=62:antenassoporte&catid=31:general&Itemid=79 [Accessed: 05-01-2019]

[4] Reuse of the Mobile Telephone System. Basic Antennas. Available from: http://antenasupv.blogspot.com/2008/07/reutilizacin-del-sitema-de-telefonia.html [Accessed: 15-10-2018]

[5] Chapter 5 Propagation Models. Available from: http://catarina.udlap.mx/u_dl_a/tales/documentos/lem/trevino_c_jt/capitulo5.pdf [Accessed: 15-10-2018]

[6] National Communications Commission (CNC), National Communications Agency (ENACOM) Resolution N° 3690/2004

[7] Azpurua M, Teixeira K, Bolívar G, Aguilar R. Evaluation of exposure to radio frequency electromagnetic fields in urban environments using spatial analysis techniques based on geographical information systems. In: Proceedings of the IV International Event of Applied Electromagnetism CNEA 2011; Santiago de Cuba March 15-18, 2011

[8] Rios Solar J. Study of non-ionizing radiations for a base station Gsm 850 Mhz located in the Private University Antenor Orrego De Trujillo [thesis]. Trujillo: Perú Private University Antenor Orrego—UPAO; 2011

[9] Figueroa de la Cruz M, Breslin R, Narváez P. Measurements of non-ionizing radiations. In: Proceedings of the VI Working Meeting on Information Processing and Control; Salta Argentina; November 17 and 18, 2016

[10] Figueroa de la Cruz M, Narvaez P, Breslin R, Den Herder T. Analysis of measurements of non-ionizing radiation in the city of Salta from the UCASAL. No. of project 123/14 [thesis]. Salta Argentina Catholic University of Salta; 2016

[11] Definition—Definition and Etymology of Words. Available from: https://definiciona.com/inmision [Accessed: 27-10-2018]

[12] Oxford Dictionary. Available from: https://es.oxforddictionaries.com/definicion/aglutinacion [Accessed: 27-10-2018]

[13] Linear Technology—LTC5533 Schottky Peak Detector. Available from: https://www.analog.com/media/en/technical-documentation/data-sheets/5534fc.pdf [Accessed: 17-10-2018]

[14] Katrin K. Surface-enhanced Raman scattering. Physics Today. 2007. DOI: 10.1007/11663898

[15] Oldenburg SJ, Hale GD, Jackson JB, Halas NJ. Light scattering from dipole and quadrupole nanoshell antennas. Applied Physics Letters. 1999;**75**:1063. DOI: 10.1063/1.124597

[16] Martin M. Surface-enhanced spectroscopy. Reviews of Modern Physics. 1985;**57**:783. DOI: 10.1007/3-540-33567-6_1

[17] Rodriguez-Fortuño F, Marino G, Ginzburg P, O'Connor D, Martinez A, Gregory A. Near-field interference for the unidirectional excitation of electromagnetic guided modes. Science. 2013;**340**(6130):328-330. DOI: 10.1126/science.1233739

[18] RECOMMENDATION ITU-R P.526-11 Available from: https://www.itu.int/dms_pubrec/itu-r/rec/p/R-REC-P.526-11-200910-S!!PDF-S.pdf [Accessed: 27-10-2018]

Chapter 4

Application of Radiation Technology: A Novel Vaccine Approach to Induce Protective Immunity against Malaria Infection

Nikunj Tandel, Devang Trivedi, Aditi Mohan Krishnan and Sarat Kumar Dalai

Abstract

Among the numerous infectious diseases, malaria remains a major health challenge. Despite the various approaches adopted for the vector control and availability of antimalarial drugs, the success of malaria eradication is dampened by the spread of drug and insecticide resistance, unavailability of proper diagnostic treatment and successful vaccine. Among the various approaches, vaccination with the aim of developing protective immunity is the most suited, safe and reliable approach for the entire mankind. Numerous approaches are in use for vaccine development; however, they suffer from the drawbacks that immunity developed is short lived and are both species- and stage-specific. Of late, radiation sterilization has drawn the attention in the vaccine development due to its advantages over the conventional methods, and successful clinical trials of irradiated vaccines against the pathogens and tumor. Recently, a novel approach of genetically attenuated sporozoites (PfRAS, PfSPZ, PFSPZ-GA1 sporozoites vaccines) has shown promising results by generating protective immunity against the homologous and heterogenous infection in the clinical trials. Radiation techniques have also been beneficial in controlling the insects by sterility technique. In this chapter, we have recapitulated the role of radiation biology in the malaria vaccine development with its current status and future challenges associated with the development of radiation attenuated parasite vaccine.

Keywords: malaria, infectious disease, vaccine, radiation technology, immunity, genetically attenuated sporozoites

1. Introduction: basic malaria biology

Malaria, an ancient disease has been found recorded in several historical documents and the word malaria comes from the Italian mal'aria meaning spoiled air [1].

It is a vector born disease caused by the parasite, belonging to genus *Plasmodium* and out of five different species, the burden of malaria leads by *Plasmodium falciparum* followed by *Plasmodium vivax* in humans [2]. 219 million cases of malaria were reported in 2017 with the death toll of 4,35,000 and the latest world malaria report suggest the stalled in the progression of the reduction in malaria elimination across the globe [3].

Almost half of the world population is at the risk of malaria, with more than 85% cases in sub-Saharan Africa followed by South-East Asia, West Pacific and others [3]. The population groups are at higher risk of malaria, and developing fatal severity include infants, children under 5 years of age, pregnant women, immunocompromised patients, non-immune migrants, mobile populations and travelers [4, 5]. In areas where malaria transmission rate is high, among these, children under 5 years are particularly receptive to infection, Illness, and death; more than 2/3 (70%) of all malaria deaths occur in this age group [4]. The number of under age of 5 years for malaria deaths has declined from 4,40,000 in 2010 to 2,85,000 in 2016, nevertheless, it remains as a leading reason for the death under 5 years children, taking the life of a child every 2 minutes [3].

1.1 Malaria life cycle

Plasmodium spp., a Causative parasite for malaria exhibits a complex life cycle that switches between *Anopheles* mosquitoes (vector), and invertebrate hosts that form unique zoite formation to evade different cell types at specific stages [6, 7]. It comprises of three different types, two of these (asexual stage) in vertebrate which are exo-erythrocytic cycle (liver/asymptomatic stage) and erythrocytic cycle (blood/symptomatic stage). And the third phase (sexual stage) is in the mosquito that is sporogenic cycle (infective stage) [7].

During the probing for the blood meal using its proboscis on the host, the infected female mosquito injects sporozoites (SPZs) within its saliva into dermis of uninfected human [8, 9] which contains vasodilators and anticoagulants that facilitate the ingestion of blood [10]. Most of the parasites reside in the skin between 1 and 6 hours [7, 8]. Many fail to migrate to the lymph nodes (LNs) and approximately 20% of them migrate directly into the skin-draining LNs through the lymph [11, 12]. It triggers the induction or modulation of the host immune system followed by an antiparasitic immune response [13, 14]. On the other hand, a small proportion of SPZs randomly transverse to the nearest blood vessel [12]. After crossing the endothelial barrier of the skin, the SPZs enter to the circulation and reached to the liver which is critical for the infection cycle as the development of merozoites occurs [15–18].

The traversal activity of SPZs through different host cells and molecular events that underpin the transformation of SPZs into merozoites via erythrocyte invasion involves the expression of thousands of proteins [7]. These SPZs migrate via blood flow to the liver retained by hepatic stellate cells (HSC) and glide, further using the Kupffer cells as a shield, traverse the liver sinusoidal endothelial cells (LESC) barrier in a sinusoids [8]. It has been moving in several hepatocytes through the space of Disse by the thrombospondin related anonymous protein (TRAP) and circumsporozoite protein (CSP) interactions before actually infecting one of them to form a parasitophorous vacuole, that aids in protecting from the host's immune system and provides nutrients for the SPZs to replicate and differentiate into merozoites [8, 10]. The healthy or successful sporozoites release up to 10,000 to more than 30 to 40,000 merozoites per hepatocyte into the bloodstream within parasite-filled vesicles called merosomes through a process of budding which use the host-cell membrane to escape the host immune system (**Figure 1**) [19].

The merozoites release in the hepatic circulation that invades erythrocytes in a short duration via dynamic and multi-step process that includes pre-invasion, active invasion, and echinocytosis [21] through the involvement of proteins named merozoite surface protein (MSP) [7]. The blood stage is symptomatic stage that develops once the erythrocytic cycle produces a parasitemia above a certain threshold approximately 50–100 parasites/μL (microscopy). The merozoites released into the blood reinvade the new red blood cells (RBCs) and continue to replicate or in some instance, they differentiate into male and female gametocytes [22, 23]. Mature gametocytes travel throughout the body and deposited mainly in skin capillaries from which they are taken up by the mosquito for the next blood meal. Once they reach to the mosquito gut, male gametocyte produces eight microgametes with three rounds of mitosis, meanwhile female gametocyte mature in macrogametes in the gut of the mosquito [24]. The fusion of the male and female gametocyte forms a diploid zygote that elongates into an ookinete; which exits the epithelium through the lumen of the gut as an oocyst and undergoes cycles of replication and forms sporozoites. It moves from the abdomen and resides into the

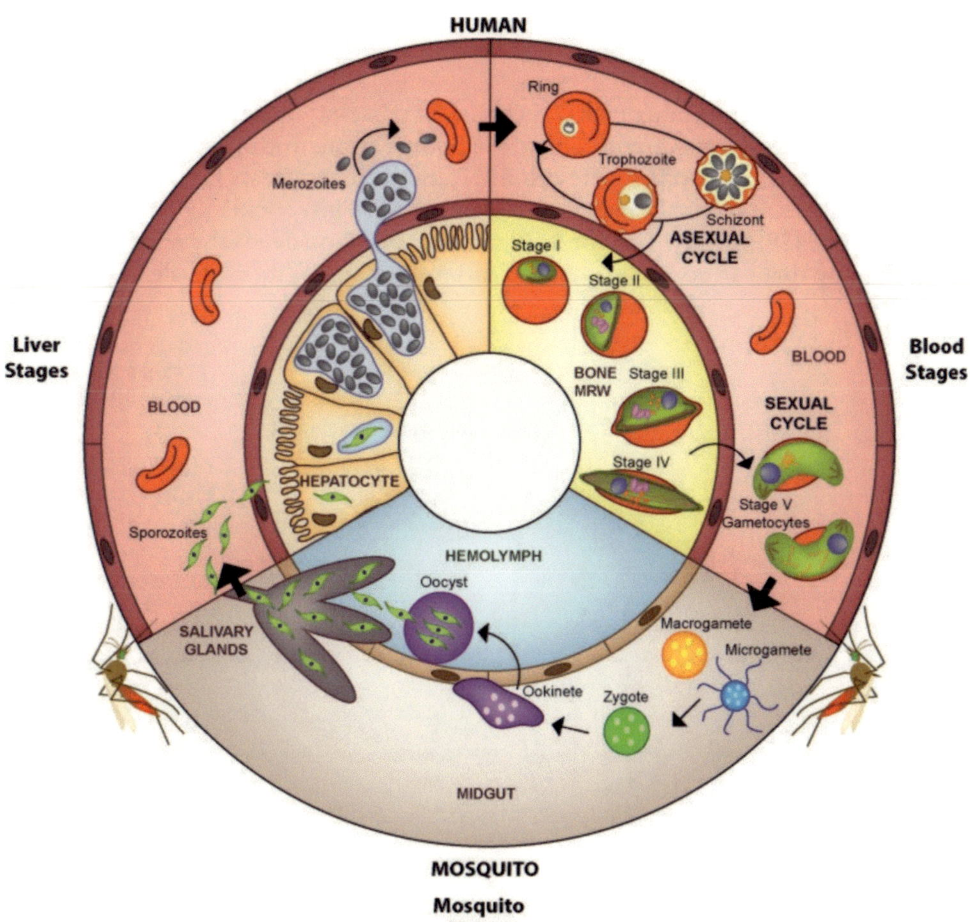

Figure 1.
The basic malaria life cycle of Plasmodium falciparum (adapted from [20] with permission).

salivary gland followed by the next blood meal of mosquito during which it will inject the sporozoites to the healthy host [24].

1.2 The liver stage infection of malaria parasites: yet hold promises

The liver is the most important organ and known for its role in blood purification, detoxification of chemicals and metabolizes of drugs, role in the immune system and homeostasis [25]. This unique architecture of liver favors the interaction between leukocytes and hepatic cells and intra-hepatic recruitment of T cells which recognize their cognate antigen within their vicinity [26]. Among the multi-stage life cycle, liver/exoerythrocytic stage of the malaria parasite is least known due to several reasons, however recent developments of humanized mice have revealed the importance of liver stage [2]. All the same, liver stage parasites in murine and humans evade the immune clearance despite the presence of antigen presenting cells (APCs) [8]. Cytotoxic T cell (CD 8^+ T cell) requires to kill the intracellular parasite infected cell as they are intracellular organism reside in the hepatocytes. Therefore long lasting immunity, reduce/remove the parasitic load in asymptomatic exoerythrocytic stage and to prevent their liver to blood stage transitions are the main concern for any malarial parasitologist to develop and improve the vaccine strategies.

1.3 Current status of antimalarial therapy and standing of vaccine initiative

To prevent and eradicate malaria, prophylaxis that has no adverse effect and versatile for all is an urgent requirement. Till now various antimalarial therapy of drugs, insecticide-treated bed nets and different diagnostic tests [27] are in use, however the success of malaria infection halted by the insecticide [28, 29] and parasite resistance [28, 30, 31]. This life-threatening condition needs the novel antimalarial therapy that can control parasite at different stages of its life cycle (**Figure 2**).

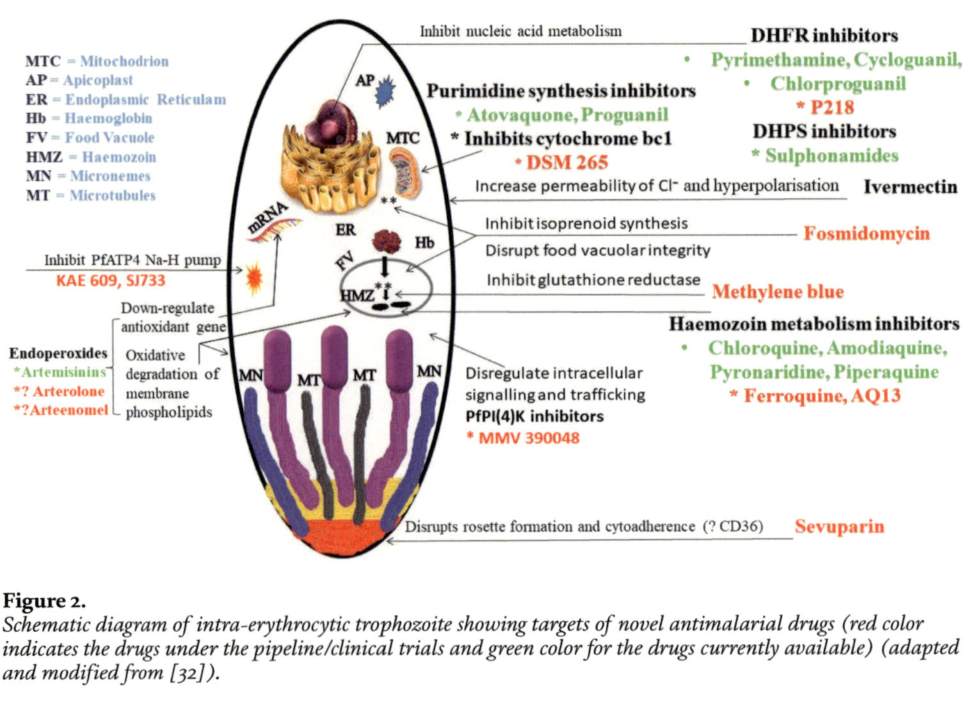

Figure 2.
Schematic diagram of intra-erythrocytic trophozoite showing targets of novel antimalarial drugs (red color indicates the drugs under the pipeline/clinical trials and green color for the drugs currently available) (adapted and modified from [32]).

Nowadays, a novel approach of combinational drug therapies are in use such as Artemisinin based combinatorial therapy (ACT) and others though, the success rate is low or restricted to the area due to the diversity among the parasite [33]. Despite the availability of drugs from natural resources or synthetic compound, the parasite still surpasses and escape the host immune system as well as dampen the effect of the drugs [34]. Also, the recent finding has alarmingly shown that parasite has been gaining resistance against artemisinin and its derivatives [31, 35]. In this bleak condition, the malaria vaccine is urgent and indeed. Different approaches are in practice for the development of malaria vaccine based on immunogenic peptides, usage of mosquito and/or parasite antigens, usage of adjutants, radiation or genetically modified sporozoites and so on [36]. Different experimental evidences shed the light on the importance of novel vaccine approaches (attenuated whole sporozoites) which strongly confirms the role of APCs resides in the liver capable to recruit CD 8+ T cells together with the cytokines (IFN-γ and TNF-α) required for the sterile protection [21, 37–43]. Despite the sterile immunity last longs for the 6–9 months, a recent study confirmed the usage of the intermittent challenge of radiation attenuated sporozoites in mice stay long lasting up to 18 months and give rise to sterile immunity [44]. The approach of radiation for protective immunity has gained attention in the scientific community to eradicate the morbidity and mortality of malaria infection.

2. Radiation and its impact on health

Radiation is a form of energy which travels and transmits or emits from its source as a wave (ionization radiation) or in the form of electron particles (non-ionizing radiation) [45]. The broad range of electromagnetic spectrum consists of harmless radio and microwaves, sunlight includes longer (infrared) and shorter (ultraviolet) wavelength and finally the higher energy specific wavelength of X-rays and gamma rays which exist the electrons from the atoms through the ionization process [45, 46]. During the process of radioactive decay there are mainly three types of ionizing radiation emitted; alpha, beta and gamma rays. Other than this, X-rays can be occurred naturally or produced by machines [47]. The major difference between this ionizing radiation (X-rays and gamma rays) is the photon's energy of individual rather than the energy of the total dose of the radiation alongside their source of origin [48].

As the entire world is in the vicinity of the radioactive environment, we all, more or less, are exposed to a certain level of background radiation [49]. It has been reported that more than 80% of human-radiation doses are uncontrolled which are mainly consist of the natural sources, terrestrial and exposure through inhalation or intake of the radiation (mainly through medication) [50].

There are more than 60 natural radionuclides are found in the environment and no place on the earth is without spontaneous radiation activity. It is mainly observed in the soils and rocks (uranium and thorium decay), in water sources such as naturally occurred lacks, rivers and oceans and Human-made buildings and homes [51]. Also, the effect of gamma radiation on natural radioactivity and their associated exposures mainly depends upon the geological and geographical location together with their appearance at different levels in the soil of the respective the region [50]. Overall, we received an average dose of 2.4 mSv/year (mSv stands for millisieverts, one-thousandth part of a sievert, an SI unit and currently used in the radiation protection standards) from the background radiation which may vary from 1 to 10 mSv/year and rely on the location of the region; however it can exceed above the 50 mSv/year. Therefore, if an average of 2 mSv/year of the radiation dose exposed to an individual, the person at the age of 80 years will be accumulated almost 160 mSv radiation originated from the natural sources [50].

2.1 Radiation biology

The term radiation biology came across into the picture during 1963 when Bergonie and Tribondeau assumed and stated that the immature, undifferentiated and continuously dividing cells are more prone towards the radiation with compare to the cells which are fully matured, differentiated and not actively participated in the cell division process [52]. Therefore, the radiosensitive cells such as stem cells, the stratum basal of the skin and stomach mucosa; continuously experiencing cell division (mitosis) and exhibits certain effects after the exposure towards the ionizing radiation resulting into the cell death or cell injury. Contradictory, the radioresistant cells such as neurons which never divides or do it very slowly show less inclined towards the cell injury or death after the radiation exposure [53]. The experiments carried out on fruit flies and mice have shown the effects of radiation as mutation was occurred however, it was notified that all the mutation were similar to the spontaneously generated one. Also, it was linked to the dose and exposure rate of the respective ionizing radiation [54].

2.1.1 Interaction between radiation and human cells

The interaction of human cells and radiation is just the likelihood and therefore, any permanent damage occurs to the tissues is not due to the facing off them each other during the cellular repair mechanism [53]. The processes of energy deposition have not any signature pattern (very rapid, 10^{-18} s) and the interaction takes place at the cellular level affect the organ as well as the entire system [45]. Alongside, there is no established cellular damage associated with the radiation, and heat, chemical or physical damage is also accounted for the same. The destruction occurs as a result of radiation towards the cells has the latent period followed by observable responses. The latent period remains for a prolonged time in case of low radiation whereas, it accounts for minutes to hours for the higher dose of radiation and the radiation biology entirely rely on these basic principles [53, 55].

The interaction of ionizing radiation with the cells, possibly occur through any of the two ways; direct interaction within the cells hit the macromolecules (proteins and DNA) resulting into the death of the cells or the mutation of the DNA. This mechanism can happen during the higher doses of the radiation as the cellular repair mechanism is tightly regulated [53]. Another one is the indirect pathway where the radiation energy is trapped inside the cellular compartment and interacts with the water rather than macromolecules followed by hydrolysis of water produced the free radicals inside the cells [53, 56]. This will lead to the loss of the important enzymes resulting in the cell death or the mutation. As the result of an interaction, there are mainly three types of cellular injury arises: (1) delayed division, (2) failure of the reproduction and (3) death during the interphase of the cell cycle [53].

The intensity of the damage occurs inside the cells rely upon various parameters including the types and its source, duration, doses, exposure and its energy [57]. The knowledge about the risk of the radiation has been studied, documented and referred well from the survivor of the atomic bombs at Hiroshima and Nagasaki in the Japan during the Second World War. Also, the additional inputs are given by the studies conducted on the radiation industry workers [45].

2.2 Radiation therapy: application in medical research

It was during the 1895, when Wilhem Roentgen has invented the X-rays to see the things visibly inside the body without surgery. This historic discovery has transformed the medical field and currently, it has been widely used to diagnose the injuries and diseases [45]. At present, more than 50% of our exposure contributed to the medical sources lead by X-rays and CT scan. Nowadays, understanding the radiation risk and their effects at molecular, cellular and organ level is mainly

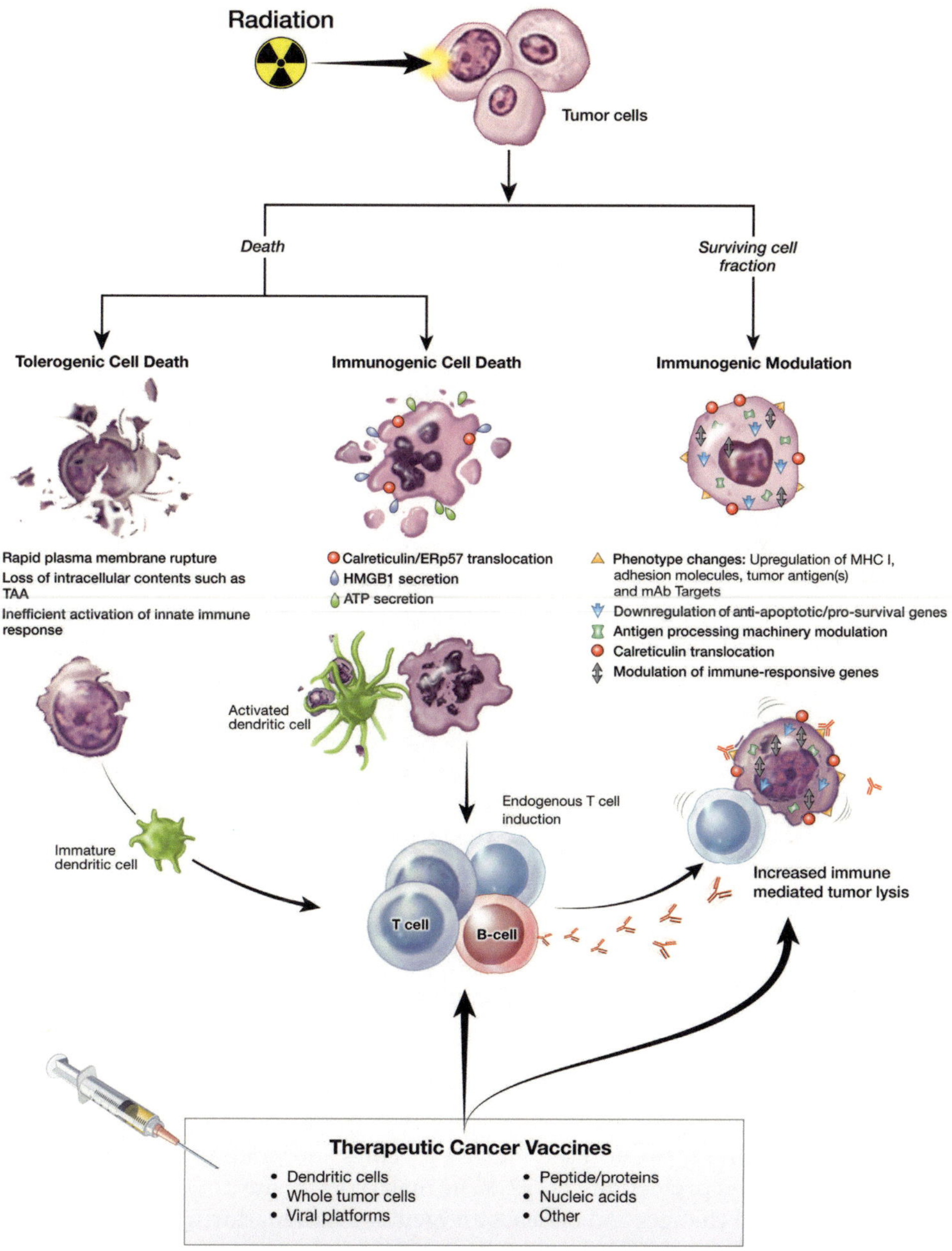

Figure 3.
Radiation therapy and their different consequences can be used for antitumor combination therapy for cancer therapeutics (adapted from [62] with permission).

studied for the diseases. This will bridge the gap and guide the health physicist to determine the safety level of the radiation to use in the medical, industrial and scientific world for the betterment of the mankind [45, 49, 58].

Despite the hazardous effects of radiation on the cells, if the cell death mechanism is instinct or properly targeted then it will be the achievement in the medical field for the cancer patients where it is using as radiation therapy [59]. It still remains as the preliminary cancer treatment and more than 50% of all the cancer patients received radiation therapy and stands for 40% of the curative treatment [60]. It is believed that it can evoke the tumor-specific immune response to destroy the tumor cells as well also travel to the site of the disease [61].

Recent research confirms the potential usage of radiation therapy in cancer therapeutics by converting tumor to favorable condition and makes them immunostimulatory milieu [62]. Experiential data suggest that the combination of radiotherapy alongside the immunotherapeutic agents can produce the synergistic effects (**Figure 3**) [61, 62]. From all the above-stated information, it has become clear that recent developments in radiation therapy [61, 63, 64] may have proven to be the most critical part in the successful development of cancer vaccine.

2.2.1 Radiation and vaccine development

Usage of chemical and physical methods to develop a stable and safe vaccine is the gold standard method through which the pathogen will be converting into the inactive form. Despite the successful application in various medical conditions to eradicate the disease or to develop the sterile immunity; the fast, reliable and rapidly generated vaccine is required [58]. Recently, radiation sterilization is been under the process for the vaccine development which surpasses the chemical or other types of contamination. Also, it destroys the nucleic acid of the respective pathogens through penetration without disturbing the cell surface antigens. It is been in under the clinical trials for the various types of cancers [65–67] however, due to the issues of safety and other regulatory affairs it is under development at industrial level [58]. On the other side, results of current clinical trials of using the various irritated vaccines for numerous pathogens and tumors has attracted the researcher and scientific world towards the preparation of various radiation-based vaccine for various infected and non-infected diseases [61, 62, 64, 68]. The ongoing effective development of irradiated vaccines for malaria and influenza have exhibited the attainability of this approach, and have shown the promising results alongside the advantages of the radiation therapy by overcoming the limitation of existing facilities without using any sophisticated technological approach [69, 70].

3. Role of radiation therapy in malaria vaccine: developments and challenges

Malaria vaccine development is an active research area with enormous challenges. An effective vaccine for *P. falciparum* is needed in malaria-endemic populations; however, none of the licensed malaria vaccines and candidates consistently produced long-lived protection [71, 72]. The malarial parasite undergoes continuous morphological changes and displays antigenic variations during the entire life cycle in both the host [73]. As a result, parasite evades the protective immune responses of the host and long-term protective immunity is not observed in malaria-infected individuals [74, 75].

Plasmodium falciparum has a multi-stage life cycle and a large 23 Mb genome expressing 5268 putative proteins. Many of these proteins exhibit allelic variation between species or antigenic polymorphism typically at sites recognized by antibody or T cell responses [76]. To date, only a handful of the nearly 5300 potential target antigens expressed by *P. falciparum*, representing less than 0.3 percent of the genome, have been pursued as vaccine targets [76, 77]. From the host perspective, malaria is a chronic infection and the *Plasmodium* parasite is capable of evading or modulating the host immune response. Based on the life cycle of the malaria parasite and the process of infection, malaria vaccines are divided into four potential target groups; interruption of human to mosquito transmission (parasite sexual and mosquito stages), inhibition of clinical consequences (asexual blood stage), prevention of mosquito to human transmission, and pre-erythrocytic infections (sporozoite [SPZ]/liver stages) [78]. As per the recent advancement in the malaria vaccine development mainly three types of vaccine candidate targeting different stages of malaria parasite have been intensively investigated named as pre-erythrocytic vaccines, blood-stage vaccines and transmission-blocking vaccines [79, 80].

3.1 Pre-erythrocytic vaccines

The pre-erythrocytic stage does not cause clinical disease, and there is no convincing evidence for naturally acquired protective immunity to this stage in individuals living in malaria-endemic areas [81]. Thus, this stage would appear to be an unattractive vaccine target. Nonetheless, the most advanced vaccine in development is a protein expressed at this stage that covers the parasite surface, the circumsporozoite protein (CSP). One of the most advanced anti-malaria vaccines at a clinical level is the RTS,S/AS01, a subunit vaccine consisting of the *P. falciparum* CSP fused with the hepatitis B surface antigen (HBsAg) [82–85]. Although this vaccine did not appear to elicit a $CD8^+$ T cell response, CSP-HBsAg induced a specific $CD4^+$ T cell response targeting the whole SPZs [86, 87]. The mechanism by which RTS,S confers protection against the blood-stage disease remains poorly understood. It seems that RTS,S induces protection against clinical malaria by temporarily reducing the number of merozoites emerging from the liver [69]. This may allow prolonged exposure to subclinical levels of asexual blood-stage parasites, therefore boosting the naturally acquired blood-stage immunity.

3.2 Blood stage vaccines

Blood-stage vaccines work on the principle of anti-invasion and anti-disease responses by blocking the invasion of erythrocytes by merozoites and preventing malarial disease [88]. The extensive genetic diversity of the parasite and the selective pressure are factors to be considered in the development of effective blood-stage vaccines. At present, several blood-stage antigens are in clinical trials: apical membrane antigen 1 (AMA1) [89], erythrocyte-binding antigen-175 (EBA-175) [90], glutamate-rich protein (GLURP) [91, 92], merozoite surface protein (MSP) 1 [93], MSP2 [94], MSP3 [95–97] and serine repeat antigen 5 (SERA5) [98]. All these antigens are highly expressed on the surface of merozoites. During the blood-stage vaccine development various assay such as ELISA, western blot and immunofluorescence assay, invasion inhibitory assay, antibody-dependent cellular inhibition (ADCI) assay, phagocytosis/opsonization assay, T cell-based assay and other antibody-based assays are in use for their screening of the candidate vaccine [99].

3.3 Transmission-blocking vaccines

Transmission-blocking vaccines (TBV) target antigens on gametes, zygotes and ookinetes to prevent parasite development in the mosquito midgut [88, 100]. The aim of these vaccines is to induce antibodies against the sexual-stage antigens to block the ookinete-to-oocyst transition to stop the subsequent generation of infectious sporozoites thereby acting as important tools for protection against epidemics [101, 102]. The leading vaccine candidates in this group include the *P. falciparum* ookinete surface antigens Pfs25 and Pfs28 and their *P. vivax* homologs Pvs25 and Pvs28. To improve the immunogenicity, Pfs25 was expressed as a recombinant protein that was chemically cross-linked to Exoprotein A and delivered as nanoparticles [103]. This enhanced the immunogenicity of the vaccine in mice, and it is currently undergoing Phase I trials in humans. Because the antigens are never naturally presented to the human immune system, one of the potential limitations of the TBV approach is that the absence of natural boosting following immunization might limit efficacy [88]. The vaccine would confer no protection to the vaccinated individual unless combined with an effective pre-erythrocytic or erythrocytic vaccine. Nevertheless, TBVs could be important tools for a malaria elimination and eradication program, for prevention of transmission of the disease [88].

3.4 Novel usages of radiation attenuated sporozoites (RAS) and effects on the host immunity

A recent landmark finding that set the standards for immunological protection against malaria infection was established by immunization with irradiated sporozoites [82, 104]. Because the parasite undergoes morphological changes and displays antigenic variation at each stage of infection, whole parasite vaccines have an advantage [105–107]. In the early 1940s, Russell and Mohan [104] first demonstrated that inactivated *P. gallinaceum* SPZs provided protection against challenge with infectious *P. gallinaceum*. In 1967, Nussenzweig et al. [82] reported that a killed *P. berghei* sporozoites (SPZs) vaccine was unsuccessful, but that an X-ray irradiated SPZ vaccine provided significant protection in an SPZ-challenge mouse model.

In the 1970s, researchers showed that immunizing human volunteers with bites from irradiated mosquitoes carrying *P. falciparum* SPZS (PfSPZ) or *P. vivax* SPZ (PvSPZs) provided protection against challenges with infectious SPZ [108–112]. Because infected mosquitoes cannot be used for immunizing large numbers of individuals, a team at the Vaccine Research Center, NIH developed an injectable and cryo-preserved irradiated PfSPZ vaccine that met the vaccine regulatory standards [58, 113]. The group succeeded in raising mosquitoes on an industrial scale to good manufacturing practice (GMP) levels and harvested large amounts of PfSPZ from the mosquito salivary glands. Although studies provided proof of concept that sporozoites could induce high-level immunity, as a vaccine for human use, PfRAS immunization was deemed impractical for many decades due to the complexity of administering a vaccine through natural way of mosquito bite, the requirement for a secure vivarium and a laboratory for maintaining *P. falciparum* in culture, and the perceived need for five or more immunization sessions to achieve a sufficient number of bites. Recently, it has been demonstrated that the Sanaria PfSPZ vaccine is safe, well tolerated, easily administered by syringe using a variety of routes, and can induce 100% protective efficacy against controlled human malaria infection (CHMI) when administered intravenously [109, 114].

The sterile immunity induced by RAS appears to be mediated primarily by $CD8^+$ and $CD4^+$ T cell-dependent mechanisms targeting antigens expressed by sporozoites and liver-stage parasites [115–118]. Antibodies also appear to contribute to protection. Studies in mice and humans show that immunization with RAS induces sporozoite-neutralizing antibodies that recognize the CSP, an abundant protein forming the surface coat of the sporozoites [109, 119, 120]. This finding led to the cloning of *P. falciparum* CSP and the formulation of several CSP-based sub-unit vaccines designed to induce protective antibodies [121, 122]. Although efficacy was low, subsequent development of CSP using a particle-based approach has led to the currently most advanced malaria sub-unit vaccine, RTS, S/AS01, that elicits 30% protection in young children [123] primarily mediated by anti-CSP antibodies and $CD4^+$ T cells [124, 125]. The partial efficacy of these first generation subunit vaccines suggests that a better understanding of RAS-induced protective mechanisms may provide a rationale to develop alternative or improved subunit strategies using newly discovered antigens or more potently inducing cell-mediated immunity. Despite all difficulties, the most convincing evidence that vaccination against malaria is feasible has come from experimental studies in rodents, monkeys and human subjects in which attenuated sporozoites induced sterile protective immunity.

3.5 Current status and future development: role of radiation in malaria vaccine development

The first whole sporozoites vaccine (WSV) studied in both rodents and humans were RAS and also the first to confer protection in humans when administered by mosquito bites. A major advance for vaccine development was the ability to isolate a purified, aseptic, and cryo-preserved product of RAS for clinical trials (PfSPZ vaccine). A subsequent scientific advance showed that intravenous (IV) administration of PfSPZ vaccine was required for inducing potent immunity in humans [109]. These studies provided the first evidence for using IV administration of a preventive vaccine in humans. More recent results with PfSPZ vaccine from sub-Saharan Africa reported that PfSPZ vaccine was well tolerated, safe, and easy to administer by direct venous inoculation (DVI) of healthy volunteers in the clinical trial setting. The proportion of participants with any infection from 28 days after the fifth vaccination to the end of the malaria season (20 weeks) was lower in the vaccinated group than in the control group [126]. These data suggest that WSV can provide some protection against malaria infection during intense transmission (93% infection rate among placebos) [79]. Recent progress in manufacturing whole organism vaccines has prompted several vaccinology questions: (1) Can whole sporozoite vaccines be improved? (2) What are the optimal doses, routes of administration, and adjuvants that confer sterilizing immunity? (3) Can other antigens be included? (4) Are we exploring all protective immune mechanisms? (5) Does efficacy differ in populations from endemic versus non-endemic areas? (6) What is the impact of antigen polymorphisms? [80].

The future lies in improving dosing strategy, immunogenicity enhancement, and/or alternative vaccine approach for populations in malaria-endemic regions. To halt the spreading of the malaria infection is not only the development of the vaccine against it, however; the mosquito biology equally holds the importance in understanding the malaria pathophysiology and parasite biology which may be understand more in detailed by radiation biology [127]. Combination of different approaches based on attenuated whole organism sporozoites may be a valuable tool that would help unearth host mechanisms of protection as well as surrogate measures of immune protection for malaria, thus providing crucial leads towards discovery of long-awaited vaccine [128].

4. Conclusion

Malaria is one of the most fatal diseases in the world related to humans in terms of morbidity and mortality. The asexual blood stage *P. falciparum* culture *in vitro* was successfully achieved in the early 1980s. However, various aspects related to their life cycle are unclear. The technological advancements in the field of epidemiology and entomology supported the research not only in reducing the burden but also allow scientists to go one step further in malaria parasitology by the staining the different stages of the parasites. The reports of the emergence of drug resistance in last decade against the various *Plasmodia* strains resulted in the serious condition to find out the causes. Given this bleak situation, the need to develop additional control measures, such as the malaria vaccine, is indeed. Understanding the mechanism of protective immunity to natural infection is critical in engineering an effective vaccine. The usage of radiation therapy in various medical fields has gained the attention of the malarial biologist to make the next generation vaccines for the long-lasting immunity. Various approaches through whole sporozoites vaccines may prove to be the better vaccine candidates, however, the obstacles related to their manufacturing, formulation, and availability to mankind needs further experimental validation.

Acknowledgement

Authors would like to thank the Department of Science & Technology (DST-SERB; grant number: EMR/2014/000543) and Department of Biotechnology (DBT; grant number: BT/PR7857/BRB/10/1215/2013), Govt. of India for providing financial support to SKD, and the Council of Scientific and Industrial Research (CSIR), New Delhi, Govt. of India for providing fellowship (CSIR-SRF HRDG No.: 09/1048(0009)/2018-EMR-I) to Mr. Nikunj Tandel.

Conflict of interest

The authors declare that they have no competing interests.

Author details

Nikunj Tandel, Devang Trivedi, Aditi Mohan Krishnan and Sarat Kumar Dalai*
Institute of Science, Nirma University, Ahmedabad, Gujarat, India

*Address all correspondence to: sarat.dalai@nirmauni.ac.in

References

[1] Cox FE. History of the discovery of the malaria parasites and their vectors. Parasites and Vectors. 2010;**3**(1):5

[2] Tyagi RK, Tandel N, Deshpande R, Engelman RW, Patel SD, Tyagi P. Humanized mice are instrumental to the study of *Plasmodium falciparum* infection. Frontiers in Immunology. 2018;**9**:19. Article Number: 2550

[3] WHO. World malaria report 2018: World Health Organization; 2018

[4] UNICEF, Malaria Mortality Among Children Under Five is Concentrated in Sub-Saharan Africa. UNICEF Data: Monitoring the Situation of Women and Children; 2015

[5] World Health Organization. Population Mobility and Malaria. World Health Organization, Regional Office for South-East Asia. 2017. Available from: https://apps.who.int/iris/handle/10665/255816

[6] Siciliano G, Alano P. Enlightening the malaria parasite life cycle: Bioluminescent plasmodium in fundamental and applied research. Frontiers in Microbiology. 2015;**6**:391

[7] Cowman AF, Healer J, Marapana D, Marsh K. Malaria: Biology and disease. Cell. 2016;**167**(3):610-624

[8] Bertolino P, Bowen DG. Malaria and the liver: Immunological hide-and-seek or subversion of immunity from within? Frontiers in Microbiology. 2015;**6**:41

[9] Schleicher TR, Yang J, Freudzon M, Rembisz A, Craft S, Hamilton M, et al. A mosquito salivary gland protein partially inhibits plasmodium sporozoite cell traversal and transmission. Nature Communications. 2018;**9**(1):2908

[10] Prudêncio M, Rodriguez A, Mota MM. The silent path to thousands of merozoites: The plasmodium liver stage. Nature Reviews Microbiology. 2006;**4**(11):849

[11] Amino R, Thiberge S, Martin B, Celli S, Shorte S, Frischknecht F, et al. Quantitative imaging of plasmodium transmission from mosquito to mammal. Nature Medicine. 2006;**12**(2):220

[12] Sidjanski S, Vanderberg JP. Delayed migration of plasmodium sporozoites from the mosquito bite site to the blood. The American Journal of Tropical Medicine and Hygiene. 1997;**57**(4):426-429

[13] Yamauchi LM, Coppi A, Snounou G, Sinnis P. Plasmodium sporozoites trickle out of the injection site. Cellular Microbiology. 2007;**9**(5):1215-1222

[14] Guilbride DL, Guilbride PD, Gawlinski P. Malaria's deadly secret: A skin stage. Trends in Parasitology. 2012;**28**(4):142-150

[15] Mota MM, Pradel G, Vanderberg JP, Hafalla JC, Frevert U, Nussenzweig RS1, et al. Migration of plasmodium sporozoites through cells before infection. Science. 2001;**291**(5501):141-144

[16] Ishino T, Chinzei Y, Yuda M. A plasmodium sporozoite protein with a membrane attack complex domain is required for breaching the liver sinusoidal cell layer prior to hepatocyte infection. Cellular Microbiology. 2005;7(2):199-208

[17] Ishino T, Yano K, Chinzei Y, Yuda M. Cell-passage activity is required for the malarial parasite to cross the liver sinusoidal cell layer. PLoS Biology. 2004;**2**(1):e4

[18] Frevert U, Engelmann S, Zougbédé S, Stange J, Ng B, Matuschewski K, et al.

Intravital observation of *Plasmodium berghei* sporozoite infection of the liver. PLoS Biology. 2005;**3**(6):e192

[19] Sturm A, Amino R, Van de Sand C, Regen T, Retzlaff S, Rennenberg A, et al. Manipulation of host hepatocytes by the malaria parasite for delivery into liver sinusoids. Science. 2006;**313**(5791):1287-1290

[20] Nilsson SK, Childs LM, Buckee C, Marti M. Targeting human transmission biology for malaria elimination. PLoS Pathogens. 2015;**11**(6):e1004871

[21] Weiss GE, Gilson PR, Taechalertpaisarn T, Tham WH, de Jong NW, Harvey KL, et al. Revealing the sequence and resulting cellular morphology of receptor-ligand interactions during *Plasmodium falciparum* invasion of erythrocytes. PLoS Pathogens. 2015;**11**(2):e1004670

[22] Baker DA. Malaria gametocytogenesis. Molecular and Biochemical Parasitology. 2010;**172**(2):57-65

[23] Waters AP. Epigenetic roulette in blood stream plasmodium: Gambling on sex. PLoS Pathogens. 2016;**12**(2):e1005353

[24] Phillips MA, Burrows JN, Manyando C, van Huijsduijnen RH, Van Voorhis WC, Wells TNC. Malaria. Nature Reviews. Disease Primers. 2017;**3**:17050

[25] Robinson MW, Harmon C, O'Farrelly C. Liver immunology and its role in inflammation and homeostasis. Cellular and Molecular Immunology. 2016;**13**(3):267

[26] Trefts E, Gannon M, Wasserman DH. The liver. Current Biology. 2017;**27**(21):R1147-R1151

[27] Tangpukdee N, Duangdee C, Wilairatana P, Krudsood S. Malaria diagnosis: A brief review. The Korean Journal of Parasitology. 2009;**47**(2):93

[28] WHO. Status Report on Artemisinin and ACT Efficacy. 2018. WHO (WHO/CDS/GMP/2018); 2018. p. 10

[29] WHO. Global Report on Insecticide Resistance in Malaria Vectors: 2010-2016. May 2018. p. 72. ISBN: 978-92-4-151405-7

[30] Thanh NV, Toan TQ, Cowman AF, Casey GJ, Phuc BQ, Tien NT, et al. Monitoring for *Plasmodium falciparum* drug resistance to artemisinin and artesunate in Binh Phuoc Province, Vietnam: 1998–2009. Malaria Journal. 2010;**9**(1):181

[31] Thanh NV, Thuy-Nhien N, Tuyen NTK, Tong NT, Nha-Ca NT, Quang HH, et al. Rapid decline in the susceptibility of *Plasmodium falciparum* to dihydroartemisinin–piperaquine in the south of Vietnam. Malaria Journal. 2017;**16**(1):27

[32] Ashley EA, Phyo AP. Drugs in development for malaria. Drugs. 2018;**78**(9):861-879

[33] Blasco B, Leroy D, Fidock DA. Antimalarial drug resistance: Linking *Plasmodium falciparum* parasite biology to the clinic. Nature Medicine. 2017;**23**(8):917

[34] Sinha S, Medhi B, Sehgal R. Challenges of drug-resistant malaria. Parasite. 2014;**21**:15. Article Number: 61

[35] Tyagi RK, Gleeson PJ, Arnold L, Tahar R, Prieur E, Decosterd L, et al. High-level artemisinin-resistance with quinine co-resistance emerges in *P. falciparum* malaria under in vivo artesunate pressure. BMC Medicine. 2018;**16**(1):181

[36] Hill AV. Vaccines against malaria. Philosophical Transactions of the Royal Society, B: Biological Sciences. 2011;**366**(1579):2806-2814

[37] Schofield L, Villaquiran J, Ferreira A, Schellekens H, Nussenzweig R, Nussenzweig V. γ Interferon, CD8+ T cells and antibodies required for immunity to malaria sporozoites. Nature. 1987;**330**(6149):664

[38] Krzych U, Dalai S, Zarling SN, Pichugin AV. Memory CD8 T cells specific for plasmodia liver-stage antigens maintain protracted protection against malaria. Frontiers in Immunology. 2012;**3**:370

[39] Krzych U, Schwenk R, Guebre-Xabier M, Sun P, Palmer D, White K, et al. The role of intrahepatic lymphocytes in mediating protective immunity induced by attenuated *Plasmodium berghei* sporozoites. Immunological Reviews. 2000;**174**(1):123-134

[40] Overstreet MG, Cockburn IA, Chen YC, Zavala F. Protective CD8+ T cells against plasmodium liver stages: Immunobiology of an 'unnatural'immune response. Immunological Reviews. 2008;**225**(1):272-283

[41] Good MF, Doolan DL. Malaria vaccine design: Immunological considerations. Immunity. 2010;**33**(4):555-566

[42] Obeid M, Franetich JF, Lorthiois A, Gego A, Grüner AC, Tefit M, et al. Skin-draining lymph node priming is sufficient to induce sterile immunity against pre-erythrocytic malaria. EMBO Molecular Medicine. 2013;**5**(2):250-263

[43] Patel H, Yadav N, Parmar R, Patel S, Singh AP, Shrivastava N, et al. Frequent inoculations with radiation attenuated sporozoite is essential for inducing sterile protection that correlates with a threshold level of plasmodia liver-stage specific CD8+ T cells. Cellular Immunology. 2017;**317**:48-54

[44] Krzych U, Dalai S, Ehrler L, Jobe O, Egner L. Infectious challenge of *Plasmodium berghei* γ-spz immunized mice rescues effector CD8+ T cells, thus assuring protracted protection. The Journal of Immunology. 2010;**184**(1 Suppl):52-59

[45] Donya M, Radford M, ElGuindy A, Firmin D, Yacoub MH. Radiation in medicine: Origins, risks and aspirations. Global Cardiology Science and Practice. 2015;**57**:438-448

[46] Reed AB. The history of radiation use in medicine. Journal of Vascular Surgery. 2011;**53**(1 Suppl):3s-5s

[47] Christensen DM, Iddins CJ, Sugarman SL. Ionizing radiation injuries and illnesses. Emergency Medicine Clinics. 2014;**32**(1):245-265

[48] Grover S, Kumar J. A review of the current concepts of radiation measurement and its biological effects. Indian Journal of Radiology and Imaging. 2002;**12**(1):21-32

[49] Thomas GA, Symonds P. Radiation exposure and health effects—Is it time to reassess the real consequences? Clinical Oncology. 2016;**28**(4):231-236

[50] Shahbazi-Gahrouei D, Gholami M, Setayandeh S. A review on natural background radiation. Advanced biomedical research. 2013;**2**:11

[51] Ramachandran T. Background radiation, people and the environment. International Journal of Radiation Research. 2011;**9**(2):63-76

[52] Bergonie J. De quelques resultas de la radiotherapie et essai de fixation d'une technique radionnelle. Comptes rendus de l'Académie des Sciences. 1906;**143**:983-995

[53] Bolus NE. Basic review of radiation biology and terminology. Journal

of Nuclear Medicine Technology. 2017;**45**(4):259-264

[54] Muller HJ. Artificial transmutation of the gene. Science. New York, NW, Washington: American Association for the Advancement of Science (AAAS), 1200, DC 20005. 1927;**66**(1699):84-87

[55] Seeram E, Travis EL. Radiation Protection. Philadelphia, New York: Lippincott; 1997

[56] Dowd SB, Tilson ER. Practical Radiation Protection and Applied Radiobiology. Philadelphia & United States of America: WB Saunders Company; 1999

[57] Authors of ICRP, Stewart FA, Akleyev AV, Hauer-Jensen M, Hendry JH, Kleiman NJ, et al. ICRP publication 118: ICRP statement on tissue reactions and early and late effects of radiation in normal tissues and organs–threshold doses for tissue reactions in a radiation protection context. Annals of the ICRP. 2012;**41**(1-2):1-333

[58] Seo HS. Application of radiation technology in vaccines development. Clinical and Experimental Vaccine Research. 2015;**4**(2):145-158

[59] Hendry J. Radiation biology and radiation protection. Annals of the ICRP. 2012;**41**(3-4):64-71

[60] Baskar R, Lee KA, Yeo R, Yeoh K-W. Cancer and radiation therapy: Current advances and future directions. International Journal of Medical Sciences. 2012;**9**(3):193

[61] Cadena A, Cushman T, Anderson C, Barsoumian H, Welsh J, Cortez M. Radiation and anti-cancer vaccines: A winning combination. Vaccine. 2018;**6**(1):9

[62] Garnett-Benson C, Hodge JW, Gameiro SR. Combination regimens of radiation therapy and therapeutic cancer vaccines: Mechanisms and opportunities. Seminars in Radiation Oncology. 2015;**25**(1):46-53

[63] Citrin DE. Recent developments in radiotherapy. New England Journal of Medicine. 2017;**377**(11):1065-1075

[64] Janiak MK, Wincenciak M, Cheda A, Nowosielska EM, Calabrese EJ. Cancer immunotherapy: How low-level ionizing radiation can play a key role. Cancer Immunology, Immunotherapy. 2017;**66**(7):819-832

[65] Qin L, Smith BD, Tsai H, Yaghi N, Neela P, Moake M, et al. Induction of high-titer IgG antibodies against multiple leukemia-associated antigens in CML patients with clinical responses to K562/GVAX immunotherapy. Blood Cancer Journal. 2013;**3**(9):e145

[66] De Remigis A, De Gruijl TD, Uram JN, Tzou SC, Iwama S, Talor MV, et al. Development of thyroglobulin antibodies after GVAX immunotherapy is associated with prolonged survival. International Journal of Cancer. 2015;**136**(1):127-137

[67] Le DT, Wang-Gillam A, Picozzi V, Greten TF, Crocenzi T, Springett G, et al. Safety and survival with GVAX pancreas prime and listeria monocytogenes–expressing mesothelin (CRS-207) boost vaccines for metastatic pancreatic cancer. Journal of Clinical Oncology. 2015;**33**(12):1325

[68] Li Y, Wang Z, Liu X, Tang J, Peng B, Wei Y. X-ray irradiated vaccine confers protection against pneumonia caused by *Pseudomonas aeruginosa*. Scientific Reports. 2016;**6**:18823

[69] Arama C, Troye-Blomberg M. The path of malaria vaccine development: Challenges and perspectives. Journal of Internal Medicine. 2014;**275**(5):456-466

[70] Jindal H, Bhatt B, Malik JS, Mehta B. Malaria vaccine: A step

toward elimination. Human Vaccines and Immunotherapeutics. 2014;**10**(6):1752-1754

[71] Beeson JG, Kurtovic L, Dobaño C, Opi DH, Chan JA, Feng G, et al. Challenges and strategies for developing efficacious and long-lasting malaria vaccines. Science Translational Medicine. 2019;**11**(474):eaau1458

[72] Mahmoudi S, Keshavarz H. Malaria vaccine development: The need for novel approaches: A review article. Iranian Journal of Parasitology. 2018;**13**(1):1

[73] Deitsch KW, Dzikowski R. Variant gene expression and antigenic variation by malaria parasites. Annual Review of Microbiology. 2017;**71**:625-641

[74] Rénia L, Goh YS. Malaria parasites: The great escape. Frontiers in Immunology. 2016;**7**:463

[75] Belachew EB. Immune response and evasion mechanisms of *Plasmodium falciparum* parasites. Journal of Immunology Research. 25 Mar 2018;**2018**:6529681

[76] Doolan DL. Plasmodium immunomics. International Journal for Parasitology. 2011;**41**(1):3-20

[77] Kooij TW, Janse CJ, Waters AP. Plasmodium post-genomics: Better the bug you know? Nature Reviews Microbiology. 2006;**4**(5):344

[78] Vaughan AM, Kappe SH. Malaria vaccine development: Persistent challenges. Current Opinion in Immunology. 2012;**24**(3):324-331

[79] Draper SJ, Sack BK, King CR, Nielsen CM, Rayner JC, Higgins MK, et al. Malaria vaccines: Recent advances and new horizons. Cell Host and Microbe. 2018;**24**(1):43-56

[80] Coelho CH, Doritchamou JYA, Zaidi I, Duffy PE. Advances in malaria vaccine development: Report from the 2017 malaria vaccine symposium. NPJ Vaccines. 2017;**2**:34

[81] Vaughan AM, Aly AS, Kappe SH. Malaria parasite pre-erythrocytic stage infection: Gliding and hiding. Cell Host and Microbe. 2008;**4**(3):209-218

[82] Nussenzweig R, Vanderberg J, Most H, Orton C. Protective immunity produced by the injection of x-irradiated sporozoites of *Plasmodium berghei*. Nature. 1967;**216**(5111):160

[83] Nussenzweig V, Nussenzweig RS. Development of a sporozoite malaria vaccine. The American Journal of Tropical Medicine and Hygiene. 1986;**35**(4):678-688

[84] Dame JB, Williams JL, McCutchan TF, Weber JL, Wirtz RA, Hockmeyer WT, et al. Structure of the gene encoding the immunodominant surface antigen on the sporozoite of the human malaria parasite *Plasmodium falciparum*. Science. 1984;**225**(4662):593-599

[85] Gordon D, McGovern T, Krzych U, Cohen J, Schneider I, LaChance R, et al. Safety, immunogenicity, and efficacy of a recombinantly produced *Plasmodium falciparum* circumsporozoite protein-hepatitis B surface antigen subunit vaccine. Journal of Infectious Diseases. 1995;**171**(6):1576-1585

[86] Kester KE, Cummings JF, Ofori-Anyinam O, Ockenhouse CF, Krzych U, Moris P, et al. Randomized, double-blind, phase 2a trial of falciparum malaria vaccines RTS, S/AS01B and RTS, S/AS02A in malaria-naive adults: Safety, efficacy, and immunologic associates of protection. Journal of Infectious Diseases. 2009;**200**(3):337-346

[87] Mahmoudi S, Keshavarz H. Efficacy of phase 3 trial of RTS, S/AS01 malaria vaccine: The need for an

alternative development plan. Human Vaccines and Immunotherapeutics. 2017;**13**(9):2098-2101

[88] Crompton PD, Pierce SK, Miller LH. Advances and challenges in malaria vaccine development. The Journal of Clinical Investigation. 2010;**120**(12):4168-4178

[89] Sagara I, Dicko A, Ellis RD, Fay MP, Diawara SI, Assadou MH, et al. A randomized controlled phase 2 trial of the blood stage AMA1-C1/Alhydrogel malaria vaccine in children in Mali. Vaccine. 2009;**27**(23):3090-3098

[90] El Sahly H, Patel S, Atmar R, Lanford T, Dube T, Thompson D, et al. The safety and immunogenicity of recombinant EBA 175-RII NG malaria vaccine in healthy adults living in a non-endemic area. Clinical and Vaccine Immunology. 2010;**17**(10):1552-1559

[91] Esen M, Kremsner PG, Schleucher R, Gässler M, Imoukhuede EB, Imbault N, et al. Safety and immunogenicity of GMZ2—A MSP3–GLURP fusion protein malaria vaccine candidate. Vaccine. 2009;**27**(49):6862-6868

[92] Hermsen CC, Verhage DF, Telgt DS, Teelen K, Bousema JT, Roestenberg M, et al. Glutamate-rich protein (GLURP) induces antibodies that inhibit in vitro growth of *Plasmodium falciparum* in a phase 1 malaria vaccine trial. Vaccine. 2007;**25**(15):2930-2940

[93] Ogutu BR, Apollo OJ, McKinney D, Okoth W, Siangla J, Dubovsky F, et al. Blood stage malaria vaccine eliciting high antigen-specific antibody concentrations confers no protection to young children in Western Kenya. PLoS One. 2009;**4**(3):e4708

[94] Genton B, Betuela I, Felger I, Al-Yaman F, Anders RF, Saul A, et al. A recombinant blood-stage malaria vaccine reduces *Plasmodium falciparum* density and exerts selective pressure on parasite populations in a phase 1-2b trial in Papua New Guinea. The Journal of Infectious Diseases. 2002;**185**(6):820-827

[95] Audran R, Cachat M, Lurati F, Soe S, Leroy O, Corradin G, et al. Phase I malaria vaccine trial with a long synthetic peptide derived from the merozoite surface protein 3 antigen. Infection and Immunity. 2005;**73**(12):8017-8026

[96] Sirima SB, Tiono AB, Ouédraogo A, Diarra A, Ouédraogo AL, Yaro JB, et al. Safety and immunogenicity of the malaria vaccine candidate MSP3 long synthetic peptide in 12-24 months-old Burkinabe children. PLoS One. 2009;**4**(10):e7549

[97] Druilhe P, Spertini F, Soesoe D, Corradin G, Mejia P, Singh S, et al. A malaria vaccine that elicits in humans antibodies able to kill *Plasmodium falciparum*. PLoS Medicine. 2005;**2**(11):e344

[98] Horii T, Shirai H, Jie L, Ishii KJ, Palacpac NQ, Tougan T, et al. Evidences of protection against blood-stage infection of *Plasmodium falciparum* by the novel protein vaccine SE36. Parasitology International. 2010;**59**(3):380-386

[99] Miura K. Progress and prospects for blood-stage malaria vaccines. Expert Review of Vaccines. 2016;**15**(6):765-781

[100] Carter R, Mendis KN, Miller LH, Molineaux L, Saul A. Malaria transmission-blocking vaccines—How can their development be supported? Nature Medicine. 2000;**6**(3):241

[101] Graves PM, Carters R, Burkot TR, Quakyi IA, Kumar N. Antibodies to *Plasmodium falciparum* gamete surface antigens in Papua New Guinea sera. Parasite Immunology. 1988;**10**(2):209-218

[102] Bousema J, Drakeley C, Sauerwein R. Sexual-stage antibody responses to

P. falciparum in endemic populations. Current Molecular Medicine. 2006;**6**(2):223-229

[103] Scaria PV, Chen B, Rowe CG, Jones DS, Barnafo E, Fischer ER, et al. Protein-protein conjugate nanoparticles for malaria antigen delivery and enhanced immunogenicity. PLoS One. 2017;**12**(12):e0190312

[104] Russell PF, Mohan B. The immunization of fowls against mosquito-borne *Plasmodium gallinaceum* by injections of serum and of inactivated homologous sporozoites. Journal of Experimental Medicine. 1942;**76**(5):477-495

[105] Hoffman SL, Goh LM, Luke TC, Schneider I, Le TP, Doolan DL, et al. Protection of humans against malaria by immunization with radiation-attenuated *Plasmodium falciparum* sporozoites. The Journal of Infectious Diseases. 2002;**185**(8):1155-1164

[106] Roestenberg M, McCall M, Hopman J, Wiersma J, Luty AJ, van Gemert GJ, et al. Protection against a malaria challenge by sporozoite inoculation. New England Journal of Medicine. 2009;**361**(5):468-477

[107] Roestenberg M, Teirlinck AC, McCall MB, Teelen K, Makamdop KN, Wiersma J, et al. Long-term protection against malaria after experimental sporozoite inoculation: An open-label follow-up study. The Lancet. 2011;**377**(9779):1770-1776

[108] Laurens MB, Billingsley P, Richman A, Eappen AG, Adams M, Li T, et al. Successful human infection with *P. falciparum* using three aseptic *Anopheles stephensi* mosquitoes: A new model for controlled human malaria infection. PLoS One. 2013;**8**(7):e68969

[109] Seder RA, Chang L-J, Enama ME, Zephir KL, Sarwar UN, Gordon IJ, et al. Protection against malaria by intravenous immunization with a nonreplicating sporozoite vaccine. Science. 2013;**341**(6152):1359-1365

[110] Epstein JE, Richie TL. The whole parasite, pre-erythrocytic stage approach to malaria vaccine development: A review. Current Opinion in Infectious Diseases. 2013;**26**(5):420-428

[111] Hoffman SL, Billingsley PF, James E, Richman A, Loyevsky M, Li T, et al. Development of a metabolically active, non-replicating sporozoite vaccine to prevent *Plasmodium falciparum* malaria. Human Vaccines. 2010;**6**(1):97-106

[112] Epstein JE, Tewari K, Lyke K, Sim B, Billingsley P, Laurens M, et al. Live attenuated malaria vaccine designed to protect through hepatic CD8+ T cell immunity. Science. 2011;**334**(6055):475-480

[113] Luke TC, Hoffman SL. Rationale and plans for developing a non-replicating, metabolically active, radiation-attenuated *Plasmodium falciparum* sporozoite vaccine. Journal of Experimental Biology. 2003;**206**(21):3803-3808

[114] Richie TL, Billingsley PF, Sim BKL, James ER, Chakravarty S, Epstein JE, et al. Progress with *Plasmodium falciparum* sporozoite (PfSPZ)-based malaria vaccines. Vaccine. 2015;**33**(52):7452-7461

[115] Malik A, Egan JE, Houghten RA, Sadoff JC, Hoffman SL. Human cytotoxic T lymphocytes against the *Plasmodium falciparum* circumsporozoite protein. Proceedings of the National Academy of Sciences. 1991;**88**(8):3300-3304

[116] Wizel B, Houghten R, Church P, Tine JA, Lanar DE, Gordon DM, et al. HLA-A2-restricted cytotoxic T lymphocyte responses to multiple *Plasmodium falciparum* sporozoite

surface protein 2 epitopes in sporozoite-immunized volunteers. The Journal of Immunology. 1995;**155**(2):766-775

[117] Krzych U, Lyon JA, Jareed T, Schneider I, Hollingdale MR, Gordon DM, et al. T lymphocytes from volunteers immunized with irradiated *Plasmodium falciparum* sporozoites recognize liver and blood stage malaria antigens. The Journal of Immunology. 1995;**155**(8):4072-4077

[118] Nardin EH, Herrington DA, Davis J, Levine M, Stuber D, Takacs B, et al. Conserved repetitive epitope recognized by CD4+ clones from a malaria-immunized volunteer. Science. 1989;**246**(4937):1603-1606

[119] Gwadz R, Cochrane A, Nussenzweig V, Nussenzweig R. Preliminary studies on vaccination of rhesus monkeys with irradiated sporozoites of *Plasmodium knowlesi* and characterization of surface antigens of these parasites. Bulletin of the World Health Organization. 1979;**57**(Suppl):165-173

[120] Egan JE, Hoffman SL, Haynes JD, Sadoff JC, Schneider I, Grau GE, et al. *Humoral immune responses in volunteers immunized with irradiated Plasmodium falciparum sporozoites*. The American Journal of Tropical Medicine and Hygiene. 1993;**49**(2):166-173

[121] Zavala F, Tam JP, Hollingdale MR, Cochrane AH, Quakyi I, Nussenzweig RS, et al. Rationale for development of a synthetic vaccine against *Plasmodium falciparum* malaria. Science. 1985;**228**(4706):1436-1440

[122] Ballou WR, Sherwood JA, Neva FA, Gordon DM, Wirtz RA, Wasserman GF, et al. Safety and efficacy of a recombinant DNA *Plasmodium falciparum* sporozoite vaccine. Bethesda, MD: Naval Medical Research Institute; 1987

[123] Olotu A, Fegan G, Wambua J, Nyangweso G, Awuondo KO, Leach A, et al. Four-year efficacy of RTS, S/AS01E and its interaction with malaria exposure. New England Journal of Medicine. 2013;**368**(12):1111-1120

[124] White MT, Bejon P, Olotu A, Griffin JT, Bojang K, Lusingu J, et al. A combined analysis of immunogenicity, antibody kinetics and vaccine efficacy from phase 2 trials of the RTS, S malaria vaccine. BMC Medicine. 2014;**12**(1):117

[125] Moorthy VS, Ballou WR. Immunological mechanisms underlying protection mediated by RTS, S: A review of the available data. Malaria Journal. 2009;**8**(1):312

[126] Sissoko MS, Healy SA, Katile A, Omaswa F, Zaidi I, Gabriel EE, et al. Safety and efficacy of PfSPZ vaccine against *Plasmodium falciparum* via direct venous inoculation in healthy malaria-exposed adults in Mali: A randomised, double-blind phase 1 trial. The Lancet Infectious Diseases. 2017;**17**(5):498-509

[127] Sinden R. The cell biology of malaria infection of mosquito: Advances and opportunities. Cellular Microbiology. 2015;**17**(4):451-466

[128] Karanja J, Kiboi N. Current milestones towards development of a fully deployable anti-malaria vaccine-future hope for malaria-free world: A review. Journal of Vaccines and Vaccination. 2016;**7**(332):2

Chapter 5

Diagnostic Imaging Safety and Protection: A Collective Interaction and Decision-Making Processes and Procedures toward an Effective Health Outcome

Chandra R. Makanjee

Abstract

This chapter will focus on the safety and protection in utilization of radiation and nonradiation imaging modalities within a medical encounter in health institutional context. The challenges of ease of access on the one hand regarding a referral versus accessing of these imaging services. The roles and responsibilities of the diverse range of professionals in the medical encounter are in ensuring that an effective decision-making is made at each point in time. The importance of communication, coordination, collaboration, and alignment is in ensuring that the care and safety of the patient is not compromised. Thus, the chore essence of provider is patient-centered care, that is, from the point of the initiation of the referral to the outcomes of an effective medical treatment and management plan. The role of the imaging investigation and its value is outweighing the risks versus harm through these chains of events and beyond.

Keywords: safety, interrelated, interdependent, distributed roles and responsibilities, collective decision-making

1. Introduction

Modern medicine is highly dependent upon high technological scientific equipment and practices [1] of which medical imaging, often referred to as a diagnostic test, is part of. The advances in medical imaging technology with enhanced image quality open the door for detecting previously unseen abnormalities of unknown relevance. According to Webster [2], imaging from surface anatomy to intrabody physiology enables the "medical gaze" to move deeper and deeper into body structures. Digitally acquired images also promote ease of access for the referrer [3]. However, technology has professional, social, and individual implications, as the availability of the latest diagnostic equipment is like a diagnostic invitation that could lead to the belief in the "gift" of knowing that would enable health-care professionals to make an informed decision [4]. Unfortunately, according to modern medicine and Smith [1], at times, the patient as a person can be lost from the clinician's gaze and advancement in patient care is not necessarily guaranteed [5].

Globally, an ongoing concern is about the effectiveness of radiation and nonradiation control measures. Increasing budgetary and financial constraints in health-care sectors and the growing consumerist movement demanding greater patient-initiated access to medical services are of special concern [6]. The justification for an imaging investigation referral or a nonreferral within in the chain of events leading to the ultimate diagnosis in managing the health outcomes lies beyond the traditional medical encounter. Whether the procedures involve exposure to ionizing radiation or nonionizing radiation, of importance is the benefit versus risks in conjunction with the clinical value of the referral, the justification of the imaging investigation requested and the actual conducting of the investigation and outcomes thereof in devising an effective management and treatment strategy that would benefit not only the health outcomes within a medical encounter but also the person.

The purpose of this chapter is to discuss the complexities of decision-making processes and procedures in imaging investigation utilization within the continuum of care processes and procedures to achieve a quality of health outcomes for the patient as a person. Where communication and interactions shaping the decisions are inherently distributed in nature. Apart from the temporary nature of the encounter, it is also integrated and intertwined within the biomedical, technological, and psychosocial dimensions. The assumption is to ensure the safety of the patient within the medical imaging context requiring a collective of decisions which are interrelated rather than isolated events that lead to a quality health outcome.

2. The health system and the timely access

Within the health system, diagnostic imaging services form an important component in terms of delivering quality professional service to health-care professionals and patients as the direct or indirect beneficiaries from the referral. The use of diagnostic medical imaging can be defined as "timely access to and delivery of integrated and appropriate radiological studies and interventions in a safe and responsive facility and a prompt delivery of accurately interpreted reports by capable personnel in an efficient, effective, and sustainable manner" ([7], p. 457).

Accessibility to these medical imaging services depends highly on the level of services provided at the various institutions. The effectiveness of these services is shaped through interactions between skilled health-care professionals and the patient and between multiple skilled health-care professionals who either work as individuals or as a group to make decisions regarding intra- and interinstitutional pathways of the referral and treatment to be followed for each individual patient. Accessing a timely service to certain imaging modalities is not always a linear pathway. In some instances, it may entail building up evidence through a graded referral from the most basic to the most sophisticated modalities based on clinical decisions. It could entail moving from one level of care to the next, depending on the availability of medical specialists, technologies [8]. A common experience of all professionals and patients is the financial boundaries and constraints. This correspond with Gibson's [9] description of "first layered approach of access to health services" aimed at cost-effectiveness and rational usage of available resources. Khan et al.'s [10] view is that the more expensive imaging services are not necessarily the most appropriate for a given clinical situation. Often, a patient's condition necessitates referral to a better resourced health-care institution for further management [11]. Then, in the absence of a central record, keeping system which includes imaging investigations may result in a re-referral for duplication of a diagnostic imaging investigation at the receiving health-care institution [12]. Regarding quality of care, it is important that the continuum of care is maintained.

The principles of justification and optimization that underpin medical practice also form the cornerstones of radiation protection [6]. On the one hand, according to Bernardy et al. [13], the quality of medical care brings value to both patient and provider when medical imaging investigation is justified and performed correctly. This investigation can be used for triaging to determine whether to refer a patient for further diagnostic tests [14]. If initial tests suggest the possibility of a condition, more costly or invasive tests may be ordered for confirmation on the basis of the possible differential diagnoses.

On the other hand, a failure of timely, appropriate clinical action following the test result can render the value of the entire process useless [7]. Medical practitioners are expected to employ the most efficient diagnostic strategies to prevent unnecessary referrals for diagnostic tests that could impact on time and resources [15]. A definitive diagnosis of a patient's condition is sometimes challenging and may require additional, unrewarding imaging examinations to improve the certainty of the diagnosis [16]. The use of radiological referral guidelines, an effective handoffs, and continuous professional development programs are essential to bridge this gap [17, 18].

Despite the acknowledgement in the current literature that health professional-patient interactions and decision-making processes regarding referral, diagnostic imaging investigations, interpretation, and communication of outcomes are complex phenomena, research in diagnostic investigations in general is distributed [8]. Effective control of ionizing or nonradiation exposure, inter alia by means of appropriate justification for every exposure, is a basic principle to be followed by all diagnostic imaging services. Each set of role players and each individual bring with them individual characteristics, skills, competencies, roles, and responsibilities. According to Webster [2], the medical gaze transcends not only at an individual level but also at a public, collective level of the regulation bodies. For instance, most health-care professionals belong to a central professional statutory regulatory body but are also governed by their institutional code of conduct and practice. An influencing or confounding factor is about financial incentives for the provider via payments by medical aids providing access to these services [9]. One way of overcoming these challenges is developing of practice-based case management. This is achieved by coordinated care management across the continuum of care by workflow mapping techniques and standardizing protocols and clinical pathways. And is followed by network interfaces between different subsystems (i.e., points of care). That is, an integrated case management platform is coordinated by clinicians as the patient navigates through the health-care delivery network [19].

3. The interrelated interdependent medical imaging encounter: decisions within a biotechnopsychosocial context

Decision-making in the health-care system—specifically with regard to diagnostic imaging investigations—occurs at multiple levels. Decision-making involves choosing a course of action to achieve specific outcomes and can occur at departmental and individual levels [17]. Within the ambit of diagnostic imaging, van Baalen et al. [20], in their study on the diagnosis and treatment of patients with pulmonary hypertension, refer to this kind of decision-making as being based on "distributed knowing." Distributed knowing implies a socially distributed process of shared meaning making among different health-care providers. Information is exchanged, collectively explored, and adjusted at the patient's different points of contact in a medical encounter. Within the health-care context, the ultimate predictor of the efficiency and effectiveness of the decision is measured by the well-being of the patient, hence the prominence given to patient-centered care.

3.1 The medical encounter

An illustration at an individual level during a medical encounter entails clinical decisions that are mostly governed by; either you have the disease or not, align with yes/no decisions. Often, these processes entail a series of interim decisions, guided at each stage to a diagnosis being present or not by minimizing uncertainty. This process still assumes there is an underlying dichotomous disease state (yes or no); this assumption could be inconclusive [21]. These interim decisions depending on the context of the encounter may be paternalistic, informed, shared, negotiated, and or a partnership process. However, these interactions should address the benefit versus risks in conjunction with the clinical value of the referral, aligning with the justification of the imaging investigation requested and the aligning of actual conducting of the investigation achieved through coordination by the medical imaging professional (competency) and cooperation by the patient mediated through text (patient records, quality of the order, and radiological report) and technology (optimally functioning equipment) to achieve an effective outcomes thereof in informing in the decision of devising an effective management and treatment strategy benefitting not only the health outcomes from a medical provider perspective but also the patient as a person.

Within the medical encounter regarding the diagnostic interaction revolves around how medical practitioners involve patients when collecting information. For example, according to Langalibalele et al., patients rely on referring doctors to provide information on the management aspect prior to the referral [12]. Not having records in this regard could lead to a decision of re-referral for duplication of investigation(s) at the receiving health-care institution. Physicians should encourage patients to describe their previous imaging examinations to help eliminate the duplication of imaging studies [16]. The dilemma is that the continuum of care is not disrupted. Often, a patient's condition necessitates referral to a better resourced health-care institution for further management [11]. If records are not centrally linked, it could lead to a re-referral for duplication of diagnostic imaging investigation at the receiving health-care institution [12, 22]. The electronic sharing of medical imaging data is an important element of modern health-care systems, but current infrastructure for cross-site image transfer depends on trust in third-party intermediaries [22, 23].

The referring doctor has the responsibility for the collection of all diagnostic information that justifies the requested radiological (radiography) examination, including information about previous exposures. Khan et al. [10] and others [24] state that in order to select imaging tests judiciously, the clinician must understand what each test can do and be fully knowledgeable about the limitations, also with regard to the available techniques. Malone et al. [6] and others [25] also refer to the use of referral guidelines or appropriateness criteria as a good practice in the process of justification. In the absence of written formal system, protocols related to the way in which diagnostic imaging investigation referrals intertwined with clinical pathways often result in what Croft et al. [21] like Croskerry [26] refers to the "gradient" of decision-making that parallels the degree of uncertainty associated with the wide variety of patient conditions, as well as to the challenge of the uncertainty about the diagnosis and the inability to stage the disease and make a choice on treatment and management.

Reasons for referrals vary, that is, to rule out a condition or to help the attending practitioner's referral decision-making. Primary care doctors commonly face the decision between ordering a test or adopting a period of "watchful waiting," requesting the patient to return later to follow the development of his or her symptoms [14, 21, 27]. To provide information to the secondary care specialist, or instance, refers to the use of radiographic investigation in triaging the referral to a secondary care specialist, whereas Gibson [9] refers to it as the medical legitimization of the institutionalization of the patient. Langalibalele et al. [12], highlight that the receiving doctor has to be well informed, otherwise there may be a need to "re-invent the wheel" through trial and error, which has an effect on cost and time. Confirmation of normality is often important in general practice to exclude or confirm a diagnosis. In this situation, a negative result may be as important as a positive one [28].

The quick-fix approaches and head-to-toe investigations were an easy way to diagnose and served as a form of reassurance that it was the right thing to do, a phenomenon that has also been reported in the international literature [29]. Then, the old paradigm of history taking, physical examination, and provisional clinical diagnosis is being replaced by imaging investigations [30]. Geneau et al. [8] refer to time management and patient overflow associated with less communication between patient and physician, which leads to medical uncertainty, and ultimately more referrals for investigations.

Health-care professionals are also of the opinion that patients see the referral as a curative measure to the extent of a total healing of their illnesses. Perceptions like these could lead to the use of technology as a placebo for the so-called demanding patients to pacify the desire for a referral and at the same time as an incentive to prevent comebacks, instead of trying to convince patients on clinical grounds why a referral for a diagnostic imaging investigation was not necessary [31]. Murphy [32] states that myths that confuse patients and "blur the boundaries between facts and fiction" are widely disseminated because of patients' previous encounters with imaging examinations. Then, some practitioners feel they deserved an investigation, because an expectation had been created and it could affect the relationship if the patient was denied of a referral for the investigation. Therefore, it is important that when decisions are made for referral are the benefits versus harm and risks are carefully weighed against each other and in situations. It is beneficial by both the medical practitioner and the patient work together to determine how to best address the situation. The patient gets an opportunity to deliberate, clarify what is most important to them and be guided [33].

3.2 The actual medical imaging investigation and outcomes thereof

The diagnostic imaging investigation phase starts with the interpretation of the request order. An investigation can only be justified if sufficient relevant clinical information is provided on the request form [34, 35]. Information gathering is initiated by the medical imaging professional interpretation of the request form. The minimal information radiographers routinely receive about their patients prior to taking a radiographic image has also been a finding in a study by Halkett et al. [36]: the quality of the information on the request form; in the event of a mismatch between the investigation requested and the intended investigation; could result in an incorrect investigation conducted; and in some instances, the correct investigation by getting additional information from the patients themselves or by contacting the general practitioner [34]. The medical imaging professional makes the choice to accept or reject the request if needed or to modify it and continue with the investigation. The quality of the task to be performed is assessed against the quality of the request and whether the referring medical officer's question is answered [37]. The value of medical records in planning the task at hand is access to patients' previous records and guides on what you could do.

Prior to commencing this subsection, it is important to provide a brief overview on the image formation and production and the importance thereof in terms of the inherent risks or harm versus benefits. To produce images depends on the type of modality used to acquire an image. A suitable source is required to produce the different forms of energy, such as X-rays (high-energy radiation), ultrasound (high-energy sound waves), magnetic resonance imaging (strong magnetic fields, electric field gradients, and radio waves), and radioactive substances. Within planar imaging, it would be, for instance, the voltage, current, and time which is depended on the distance and type of receptor used. For ultrasound, it would be the type of transducer used which generates the sound pulses and detects the echoes. Whereas, with MRI, it entails selecting appropriate imaging parameters like T1 and T2 and the various software available to characterize the image. To produce an image requires a suitable medium to capture these attenuated energies and convert to an analogue or digital form of a visible image on a screen to make a diagnosis. All of these depend on the ability and the capability of the imaging equipment, the competency of the operator to make a sound decision on the completeness of the investigation using sound scientific knowledge-based approach without compromising the integrity of the quality of the examination.

For example, in the case of follow-up imaging investigation, it may be modified, so with establishing the exposure technique, some factors determine the quality of the image. In the case of pathology, sometimes, the exposure technique needs to be adapted is to be consistent of the quality standard of the image produced to compare with previous images. This is governed, among other factors (e.g., the focus to film distance and positioning of the patient), by the exposure technique over which the radiographer has most control [38–40]. Precautionary measures have a positive outcome on the possible risk of exposing the patient unnecessarily to radiation. Part of obtaining an optimal quality image is the investigation protocol and procedures including the imaging parameters [41]. Established departmental quality control and assurance guidelines are essential to avoid inconsistencies in practice which may result in suboptimal imaging investigation. The patient's physical condition and capacity to cooperate in the examination must be assessed and any shortcomings must be communicated to others in the health-care team [39, 42]. Radiographers often have trouble in acquiring the desired projection if the patient is either uncooperative or immobile. The investigation is measured against the time consumed and the worth in terms of anticipated normal versus abnormal findings.

The completeness of the information required to generate a radiological diagnostic report is depended on the quality of the completeness of the request, the patient, and the accuracy of investigation performed. According to Khan et al. [10], detailed case notes and a well-conceived, ordered list of differential diagnoses are the absolute minimum to include in any imaging request to ensure that the selected imaging is warranted and to improve the accuracy of reporting. Then, clinicians should not just read the radiological report, but ought to be able to interpret the image. Misreading of images has been shown to be the most common type of clinical error [28]. According to Hardy and Barrett [43], a referral for a diagnostic investigation stems from a clinical examination, based on the clinical signs and symptoms. The provisional diagnosis can be confirmed or refuted depending on the clinician's ability to interpret the images. Therefore, in all circumstances, the decision to do a radiographic investigation should be influenced by the ability to interpret the resultant image [27]. It is the responsibility of the treating clinician to determine whether the anatomic anomaly revealed by an imaging study is related to the patient's symptoms [10]. This could be since in clinical practice "to recognize pathology in a 'sea of normals'" is quite difficult and "[t]he prevalence of pathology can contribute to a 'context or prevalence bias' in decision making" [44]. It is recommended that collaborative radiologist-medical practitioner educational efforts to help enhancing medical practitioners' knowledge could be useful. Another option could be use of decision aids [45].

3.3 Patient autonomy

Patient-centered care puts the patient as partner and collaborator in the diagnosis and management of his or her own health conditions. Patient-centered care is intricately linked with notions of shared decision-making and the patient's active participation in all processes of the medical encounter, ranging from the provisional diagnosis to the choice of diagnostic investigations (including diagnostic imaging), the discussion of provisional findings, and the design of a management plan [46–53]. In the doctor-patient consultation, there has been a steady shift from paternalism toward a focus on the needs and the multiple voices of providers and the autonomy of the patient as being at the center of his or her own care [54]. When information is passed on, an individual's language preferences and level of literacy should be considered [50], including the ability to negotiate and coordinate care [47]. One way of potentially solving language discordance problems and reducing disparities in care is to provide language interpreters [55] for the patient to understand what is going on and could engage effectively [56] to bridge this gap by use of comprehensive language and modify some of the terminology through metaphors [57]. In communicating the events frequently, little or no mention is made of radiation risk [6, 58]. The argument is the complex and specialized nature of the units used to quantify radiation exposure, which is not conducive to effective communication with the public and even with health professionals. Patients have the right to know of the radiation risk and it is the duty of health professionals to inform them [6]. This contributes to empowering patients to make informed decisions especially in the case of high-dose procedures, where open discussion and shared decision-making would facilitate the process. Radiologists and their registrars had an expectation that it was the attending medical officers' responsibility to communicate risks and benefits. This type of consultation is also dependent on the institutional culture which promotes active participation or where patients are expected to behave like "good" patients and passively accept services and attention allocated to them [9].

Another important factor is the potential influence of consumer awareness and demand on the patterns of utilization of diagnostic imaging services. Patients may demand imaging procedures for various reasons: they may have acquired information from the print or electronic media or by word of mouth; or they could believe that they should receive specific imaging services for particular symptoms, based on their past experience. Most patients are not financially liable for imaging services received. If the physician is reluctant to refer them for diagnostic imaging, they may interpret it as insensitivity on the part of the physician who withholds procedures that they are entitled to. Furthermore, many patients have little understanding of indications for or benefits of imaging procedures and the cost involved. Radiation doses and their associated risks, and the protection procedures in place are also poorly understood [6, 8, 16, 59, 60]. One of the gaps in the current knowledge on the functioning of the health system is the extent to which patients are aware of their rights regarding participation in the planning of their treatment and diagnostic processes, including knowledge of radiation risks. According to Geneau et al. [8], patient demands influence physicians' behavior. Espeland and Baerheim [29] found that medical practitioners complied with strong wishes from patients in cases where the clinical indication for radiographic investigation was in doubt, little else could be done, the consultation was difficult, or time was scarce or out of moral obligation. This created a false sense hope that something could and had been done. A choice was made to refer the patient without the knowledge of what the patient's desires were in terms of monetary incentives. The types of misconceptions of passive demands or expectations may have been created as a result of the "quick fix" approach, for instance, like, let us rather do the X-ray, it is easier. Khan et al. [10] blame physicians for contributing to the idea that an accurate diagnosis is only possible with the aid of an image. The above view corresponds with Balaqué and Cedraschi's [61] contention that "[p]atients tend to consider technological investigations as more trustworthy than the clinical examination." According to Borgen et al. [62], "normal findings will reassure the patient" (p. 197) and treatment, gives patients the feeling of being taken seriously.

4. The professional confined spaces yet inherently multiprofessional

According to Mørk et al. [63], shared practice by its very nature creates boundaries (p. 14), and experiences may be modified and extended in the light of experiences in their discipline fields. For example, boundary blurring between practices of radiologists and surgeons necessarily evokes conflicts and that each group wants to claim ownership to the treatment and to the eligible patients [63]. Powell and Davies [65] describe professional territory as "the differences in professional identities and core beliefs" and the significant impact of professional identities and boundaries on how individual health-care providers from the same or different profession work with each other—something that has implications for the care that patients receive. They also refer to the radiology profession seeking to lay claim to particular fields of knowledge and to assert their jurisdiction over particular tasks. In a study by Johansen and Brodersen [66], the fear of losing demarcations with regard to resources and organizational quality is also a concern between medical specialist professionals and radiographers where tasks were taken over or shared. Similar to Stephens and Carmeli [64], Hilligoss [67] studied that existing personal relationships, differing levels of experience, formal power structures, and hierarchies have numerous effects on quality of care or services. Lack of information and understanding of professional roles and responsibilities, meaningful communication, and relationships are also reported in the literature [1, 47]. The radiographer's role in theater is confined to a task of taking images, but usually there is no direct cooperation between, for example, operation nurses and radiographers [63]. The power of relations in the hierarchies leaves very little space for radiographers to participate in decision-making processes dominated by other professions. Medical practitioners see the completion of the request form as a medium of instruction to perform a job. A radiologist should—before accepting an examination request—be aware of the clinical condition of the patient and the preceding examinations, to be able to make appropriate decisions [35] in the event of not having the full picture of the interactions that had taken place prior to the referral. They were respectful of the doctor-patient relationship and did not want to be the confounders, which could often result in conflict. Lewis et al. refer to unethical situations relating to the justification of radiographic examinations and radiographers' feelings of uncertainty regarding their legal and moral responsibilities [68]. Olivier et al. see radiographers as often being in the forefront where patients want to know from them what is wrong [69]. They emphasize the importance of finding the appropriate words that will keep them within their professional boundary. If the radiographer is not able to disclose results to patients, it does not afford them much professional autonomy in their working environment [50]. Often the patient is at risk of not been communicated too.

The biomedical and psychosocial worlds cannot be treated as isolated components [58]. These authors refer to the different languages and cultures of these worlds—different ways of knowing—that both contribute to the establishment of overall care of the patients, inter alia about efficient use of resources, the quality of services, and provider and patient satisfaction. These authors refer to the predictability existing in the biomedical world with its focus on anatomy and physiology with a view to diagnose and institute effective treatment to bring the human body back to its "normal state." This is in contrast with the greater unpredictability and complexity inherent in the psychosocial nature where cognition, emotion, and behavior function more at a normative level needed for adaptability and flexibility. **Figure 1** represents an attempt to interpret the biopsychosocial interactions within "a bigger picture" [58] that portray the nonstatic nature of health-care provision with ever-changing and emerging ways of treatment and health management interaction with recent technological evidence [51].

There is also a growing body of knowledge on the interpretation of biotechno interactions in the ambit of medical technoscience; for example, relations between the analog and digital technological worlds of communication and interactive processes, and the fusion between diagnostic and therapeutic work [47, 63, 71].

Figure 2 depicts the role of information and interpretation in diagnostic imaging decision-making. At each point of contact, four interrelating activities with the focus on information take place: information gathering, information verification, information processing, and information exchange. All these activities also play a role in the final transformation of information, that is, integration and interpretation needed for completing the assessment-treatment-expected outcome sequence [70]. This is a continuous cyclical circular process that plays itself out throughout the patient's journey.

The decision-making and diagnosis processes are also characterized by information and knowledge inputs and outputs that contribute to an awareness of the bigger picture and which Wilson [72] considers as a basis for decision-making processes that could lead to improved operational, economic, or clinical benefit

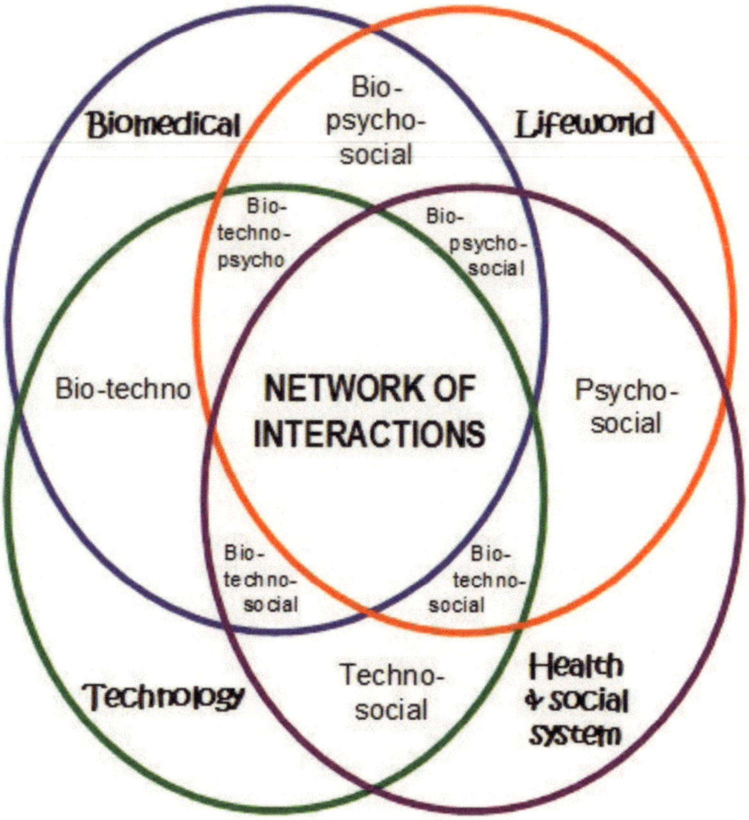

Figure 1.
The biotechnopsychosocial network of interactions.

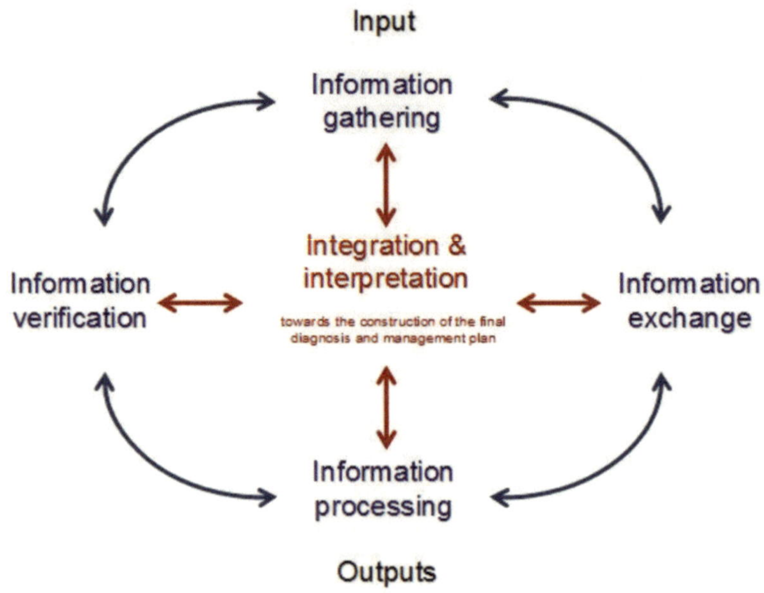

Figure 2.
The role of information and interpretation in diagnostic imaging decision-making.

[73]. According to Paul and Reddy, interpretation draws on mediation tools and embedded contexts such as work practices, cultures, organization structures, and interpersonal relations [74].

To construct and reconstruct a diagnosis and management plan to improve patient outcomes, information in the form of empirical evidence—the "right" piece of information—needs to be gathered during a medical consultation and by means of diagnostic tests [75, 76]. Pivotal questions are "what information to gather?; which diagnostic test to perform?; how to interpret and integrate this information to draw diagnostic conclusions?" ([75], pp. 26–27).

Medical imaging procedures whether diagnostic or interventional, for instance, draw the following together: the system; diverse but interconnected communities of practice; the patient; technology; drugs; clinical interventions; and many other elements [77]. All of them are interconnected and interrelated, "yet each irreducible to the other."

Regarding diagnostic imaging, Murphy [78] distinguishes between "hard technology" (equipment) designed to diagnose and treat disease and "soft technology" that includes the social interactions of radiographers with patients and other health-care professionals. The radiographer acts as an interface between the patient and biomedical health, technology, and humane health care, referred to as "technology-in-practice" [79]. The coordinating power of health technologies is concerned with how technologies can bring together or break apart the pragmatic worlds as social actors navigate in our everyday lives [80]. Reeves and Decker [81] refer to the "technology-human dualism" with which the profession of radiography is faced because of the short encounter and once-off interactions with patients discouraging emotional investment. Within the imaging context entails an acknowledgement of the situation-specific encounter between individual patients and health-care professionals where patient needs and desires do not come to the forefront instead the anatomy and physiology [70].

Some of these combined terminologies referred to above are well described in the literature, whereas others need further exploration in future in terms of the feasibility of their application and their relations to quality of patient care, patient and provider satisfaction, and/or efficient use of resources [70]. The ever-shifting boundaries between different worlds in which health-care providers and patients are situated and the interconnectedness between role players also has particular implications for the formation of professional boundaries, and identities illustrating the temporariness of encounters within the continuum of the care is the challenge of ensuring the safety of the patient as a person.

5. Conclusions

In conclusion, the building blocks for deciding on the most appropriate investigation of choice with minimal risk and optimal benefit in terms of management and treatment strategies are highly dependent on the organizational structure, its institutional members, the quality of the referral, the investigation itself, and the outcomes thereof. The process evolves as the events unfold, based on the actions that are taken and highly dependent on who is communicating with whom in that institution or with a referral institution and what is communicated with whom. The processes and interactions are also more dependent on what patients present with and how patients present their condition. It calls for a risk-centered approach to many syndromes and chronic conditions [6] and parallels proposals that public health should be organized around achievable outcomes rather than disease categories. Such a framework shifts the focus of clinical practice to improving outcomes for patients in their total biological, psychological, and social environment and away from an exclusive and narrow focus on underlying disease as the determinant of outcome. The underlying "disease" is often a continuous distribution of probability for future health states [21]. This encounter entails the patient together with a diverse range of health-care professionals to collectively align their decisions in ensuring that the safety and care of the patient as a person is not compromised in the delivery of health-care services.

Conflict of interest

None.

Author details

Chandra R. Makanjee
Department of Medical Radiation Sciences, University of Canberra, Bruce, ACT, Australia

*Address all correspondence to: chandra.makanjee@canberra.edu.au

References

[1] Adams J, Smith T. Qualitative methods in radiography research: A proposed framework. Radiography. 2003;**9**(3):193-199

[2] Webster A. Innovative health technologies and the social: Redefining health, medicine and the body. Current Sociology. 2002;**50**(3):107-122

[3] Davies HE, Wathen CG, Gleeson FV. Clinical review risks of exposure to radiological imaging to minimise them. BMJ. 2011;**342**(947):589-593

[4] Kenen RH. The at-risk health status and technology: A diagnostic invitation and the "gift" of knowing. Social Science & Medicine. 1996;**42**(11):1545-1553

[5] Munn Z, Jordan Z. The patient experience of high technology medical imaging: A systematic review of the qualitative evidence. Radiography. 2011;**17**(4):323-331

[6] Malone J, Guleria R, Craven C, Horton P, Järvinen H, Mayo J, et al. Justification of diagnostic medical exposures, some practical issues: Report of an International Atomic Energy Agency Consultation. The British Journal of Radiology. 2013;**85**:523-538

[7] Pitman A, Jones DN, Stuart D, Lloydhope K, Mallitt K, O'Rourke P. Royal Australian and New Zealand College of Radiologists (RANZCR) relative value unit workload model, its limitations and the evolution to a safety, quality and performance framework. Journal of Medical Imaging and Radiation Oncology. 2009;**53**(5):450-458

[8] Geneau R, Lehoux P, Pineault R, Lamarche P. Understanding the work of general practitioners: A social science perspective on the context of medical decision making in primary care. BMC Family Practice. 2008;**9**(12):1-10

[9] Gibson D. Negotiating the new health care system in Cape Town, South Africa: Five case studies of the acutely chronically ill. Medical Anthropology Quarterly. 2001;**15**(4):515-532

[10] Khan KM, Tress BW, Hare WS, Wark JD. Treat the patient not the X-ray: Advances in diagnostic imaging do not replace the need for clinical interpretation. Journal of Sports Medicine. 1998;**8**(1):1-4

[11] Smith S, Khutoane G. Why doctors do not answer referral letters. South African Family Practice. 2009;**51**(1):64-67

[12] Langlibalele M, Maduna PMH, Mhlongo SW, Ndimande JV, Longo-Mbenza B. Quality of general practitioner referral letters to a South African tertiary hospital: Determinants of quality content and good practice. Journal of Public Health and Epidemiology. 2011;**3**(11):482-488

[13] Bernardy M, Ullrich CG, Rawson JV, Allen B, Thrall JH, Keysor KJ, et al. Strategies for managing imaging utilization. Journal of American College of Radiology. 2009;**6**(12):844-850

[14] Whiting P, Toeriena M, De Salisa I, Sternea JAC, Dieppea P, Eggerb M, et al. A review identifies and classifies reasons for ordering diagnostic tests. Journal of Clinical Epidemiology. 2007;**60**(10):981-989

[15] Kravitz RL, Callahan EJ. Patient's perceptions of omitted examinations and tests: A qualitative analysis. Journal of General Internal Medicine. 2000;**15**(1):38-45

[16] Hendee WR, Becker GJ, Borgstede JP, Bosma J, Casarella WJ, Erickson BA, et al. Addressing overutilization in medical imaging. Radiology. 2010;**257**(1):240-245

[17] Patel V, Kaufman D, Arocha J. Emerging paradigms of cognition in medical decision-making. Journal of Biomedical Informatics. 2002;**35**:52-75

[18] Rawle M, Pighills A. Prevalence of unjustified emergency department X-ray examination referrals performed in a regional Queensland hospital: A pilot study. Journal of Medical Radiation Science. 2018;**65**(3):184-191

[19] Meyer HC, Jekowsky E, Crane FG. Applying platform design to improve the integration of patient services across the continuum of care. Managing Service Quality: An International Journal. 2007;**17**(1):23-40

[20] van Baalen S, Carusi A, Sabroe I, Kiely DG. A social-technological epistemology of clinical decision-making as mediated by imaging. Journal of Evaluation in Clinical Practice. 2016;**23**:949-958

[21] Croft P, Altman DG, Deeks JJ, Dunn KM, Hay AD, Hemingway H, et al. The science of clinical practice: Disease diagnosis or patient prognosis? Evidence about "what is likely to happen" should shape clinical practice. BMC Medicine. 2015;**13**:20. DOI: 10.1186/s12916-014-0265-4

[22] Eskeland SL, Brunborg C, Rueegg CS, et al. Assessment of the effect of an Interactive Dynamic Referral Interface (IDRI) on the quality of referral letters from general practitioners to gastroenterologists: A randomised cross-over vignette trial. BMJ. 2017;7(6):e014636. DOI: 10.1136/bmjopen-2016-014636

[23] Patel V. A framework for secure and decentralized sharing of medical imaging data via blockchain consensus. Health Informatics Journal. 2018:1-14. DOI:10.1177/1460458218769699

[24] Hollingsworth TD, Duszak R, Vijayasarathi A, Mullins ME. Trainee knowledge of imaging appropriateness and safety: Results of a series of surveys from a large Academic Medical Center. Current Problems in Diagnostic Radiology. 2019;**48**:17-21

[25] Vilar-Palop J, Hernandez-Aguado I, Pastor-Valero M, et al. Appropriate use of medical imaging in two Spanish public hospitals: A cross-sectional analysis. BMJ Open. 2018;**8**:e019535. DOI: 10.1136/bmjopen-2017-019535

[26] Croskerry K. The theory and practice of clinical decision-making. The Canadian Journal of Anaesthesia. 2005;**52**(1):R1-R8

[27] Smith AN. Remote X-ray operator radiography: A case study in interprofessional rural clinic practice [doctoral thesis]. New South Wales, Australia: University of Newcastle; 2006. Available from: http:// www.nova.newcastle.edu.au/vital/access/manager/Repository/uon: 709 [Retrieved: 7 September 2007]

[28] Durham JA, Mcleod DK. A qualitative study of general practitioner access to diagnostic imaging services in the central region. The New Zealand Medical Journal. 1999;**112**(1089):211-213

[29] Espeland A, Baerheim A. Factors affecting general practitioners' decisions about plain radiography for back pain: Implications for classification of guideline barriers—A qualitative study. BMC Health Services Research. 2003;**3**(8):1-10

[30] Mendelson RM, Murray CPJ. Towards the appropriate use of diagnostic imaging. Medical Journal of Australia. 2007;**187**(1):5-6

[31] Makanjee CR, Bergh AM, Hoffmann WA. Patients' journeys through multi-level diagnostic imaging referrals. The South African Radiographer. 2015;**53**(1):11-16

[32] Murphy F. Understanding the humanistic interaction with medical imaging technology. Radiography. 2001;**7**(3):193-201

[33] Dobler CC. Quality of shared decision making in lung cancer screening: The right process with the right partners at the right time and place. Mayo Clinic Proceedings. 2017;**92**(11):1612-1616. DOI: 10.1016/j.mayocp.2017.08.010 www.mayoclinicproceedings.org

[34] Dhingsa R, Finlay DBL, Robinson GD, Liddicoat AJ. Assessment of agreement between general practitioners and radiologists as to whether a radiation exposure is justified. British Journal of Radiology. 2002;**75**(890):136-139

[35] Triantopoulou C, Tsalafoutas I, Maniatis P, Papavdis D, Raios G, Siafas I, et al. Analysis of radiological examination request forms in conjunction with justification of X-ray exposures. European Journal of Radiology. 2005;**53**(3):306-311

[36] Halkett GKB, McKay J, Shaw T. Improving students' confidence levels in communicating with patients and introducing students to the importance of history taking. Radiography. 2011;**17**(1):55-60

[37] Larsson W, Hillergård K, Lundberg N. Use your good judgment—Radiographers' knowledge in image production work. Radiography. 2009;**15**(3):11-21

[38] Lam D, Egan I, Baird M. The radiographer's impact on improving clinical decision-making, patient care and patient diagnosis: A pilot study. The Radiographer. 2004;**51**(3):133-137

[39] Mc Inerney J, Baird B. Developing critical practitioners: A review of teaching methods in the bachelor of radiography and medical imaging. Radiography. 2016:e40-e53. DOI: 10.1016/j.radi.2015.07.001

[40] Moskowitz H. I.C.U. Chest Radiology: Principles and Case Studies. 1st ed. Hoboken, NJ: John Wiley and Sons; 2010

[41] Carlton RR, Adler AM. Principles of Radiographic Imaging: An Art and A Science. 5th ed. New York: Delmar, Cengage Learning; 2013

[42] Egan I, Baird M. Optimising the diagnostic imaging process through clinical history documentation. The Radiographer. 2003;**50**(1):11-18

[43] Hardy M, Barret C. Interpretation of trauma radiographs by radiographers and nurses in the UK: A comparative study. The British Journal of Radiology. 2004;**77**(920):657-661

[44] Pusic VC, Andrews JS, Kessler DO, Teng DC, Pecaric MR, Ruzal-Shapiro C, et al. Prevalence of abnormal cases in an image bank affects the learning of radiograph interpretation. Medical Education. 2012;**46**(3):289-298

[45] Spilseth B, Ghai S, Patel NU, Taneja SS, Margolis DJ, Rosenkrantz AB. A comparison of radiologists' and urologists' opinions regarding prostate MRI reporting: Results from a survey of specialty societies. American Journal of Roentgenology. 2018;**210**(1):101-107. DOI: 10.2214/AJR.17.18241

[46] Undeland M, Malterud K. Diagnostic interaction: The patient as a source of knowledge? Scandinavian Journal of Primary Health Care. 2008;**26**(4):222-227

[47] Suter E, Arndt J, Arthur N, Parboosingh J, Taylor E, Deutschlander S. Role understanding and effective communication as core competencies for collaborative practice. Journal of Interprofessional Care. 2009;**23**(1):41-45

[48] Frank AW, Corman MK, Gish JA, Lawton P. Healer-patient interaction: New mediations in clinical relationships. In: Bourgeault I, Dingwall R, De Vries R, editors. The SAGE Handbook of Qualitative Methods in Health Research. 1st ed. London: Sage; 2010. pp. 34-52

[49] Bleakley A. Professing medical identities in the liquid world of teams. Medical Education. 2011;**45**(12):1167-1173

[50] Etheredge H. Rethinking responsibility in radiography: Some ethical issues in South Africa (opinion piece). SA Journal of Radiology. 2011:10-13

[51] Makanjee CR, Bergh AM, Hoffmann WA. Healthcare provider and patient perspectives on diagnostic imaging investigations. African Journal of Primary Health Care and Family Medicine. 2015;7:1-10

[52] Cropley S. The relationship-based care model: Evaluation of the impact on patient satisfaction, length of stay, and readmission rates. The Journal of Nursing Administration. 2012;**42**(6):333-339

[53] Dy SM, Purnell TS. Key concepts relevant to quality of complex and shared decision-making in health care: A literature review. Social Science & Medicine. 2012;**74**(4):582-587

[54] Elwyn G. Arriving at the postmodern medical consultation. Primary Care. 2005;**5**(12-13):287-291. Available from: http://www.primary-care.ch/docs/primarycare/archiv/defr/2005/2005-12/2005-12-423.PDF [Retrieved: 5 January 2009

[55] Jayadevappa R, Chattre S. Patient centred care—A conceptual model and review of the state of art. The Open Health Services and Policy Journal. 2011;**4**:15-25

[56] Henderson A. Boundaries around the 'well-informed' patient: The contribution of Schutz to inform nurses' interactions. Journal of Clinical Nursing. 2006;**15**(1):4-10

[57] Van Ravesteijn H, Van Dijk I, Darmon D, Van Der Laar F, Lucassen P. The reassuring value of diagnostic tests: A systematic review. Patient Education and Counseling. 2012;**86**(1):3-8

[58] Makanjee CR, Engel-Hills P. Ethics in diagnostic radiography in South Africa: A complex temporary encounter mediated through text and technology. In: Nortjé N, De Jongh JC, Hoffmann W, editors. African Perspectives on Ethics for Healthcare Professionals. Advancing Global Bioethics. Vol. 13. Cham: Springer; 2018. pp. 201-204

[59] Chun-sing W, Bingshenga H, Ho-kwan S, Wai-lamb W, Ka-ling Y, Tiffany CYC. A questionnaire study assessing local physicians, radiologists and interns' knowledge and practice pertaining to radiation exposure related to radiological imaging. European Journal of Radiology. 2012;**81**(3):264-268

[60] Lin EC. Radiation risk from medical imaging. Mayo Clinic Proceedings. 2010;**85**(12):1142-1146. quiz 1146

[61] Balaqué F, Cedraschi C. Radiological examination in low back pain patients: Anxiety of the patient? Anxiety of the therapist? Joint, Bone, Spine. 2006;**73**(5):508-513

[62] Borgen L, Stranden E, Espeland A. Clinicians' justification of imaging: Do radiation issues play a role. Insights Imaging. 2010;**1**(3):193-200

[63] Mørk BE, Aanestad M, Hanseth O, Grisot M. Conflicting epistemic cultures and obstacles for learning across communities of practice. Knowledge and Process Management. 2008;**15**(1):12-23

[64] Stephens JP, Carmeli A. Relational leadership and creativity: The effects of respectful engagement and caring on meaningfulness and creative work involvement. In: Hemlin S, Mumford MD, editors. Handbook of Research on Creativity and Leadership. Edward Elgar Publishing. 2017

[65] Powell AE, Davies HTO. The struggle to improve patient care in the face of professional boundaries. Social Science & Medicine. 2012;**75**(5):807-814

[66] Johansen LW, Brodersen J. Reading screening mammograms: Attitudes among radiologists and radiographers about skill mix. European Journal of Radiology. 2011;**80**(3):325-330

[67] Hilligoss PB. Patient handoffs between emergency department and inpatient physicians: A qualitative study to inform standardization of practice and organization theory [doctoral thesis]. Ann Arbor, Michigan, United States: University of Michigan; 2011. Available from: http://deepblue.lib.umich.edu/ bitstream/2027.42/86293/1/bhilligo_1.pdf [Retrieved: 7 May 2012]

[68] Lewis S, Heard R, Robinson J, White K, Poulos A. The ethical commitment of Australian radiographers: Does medical dominance create an influence? Radiography. 2008;**14**(2):90-97

[69] Olivier L, Leclère J, Dolbeault S, Neuenschwander S. Doctor-patient relationship in oncologic radiology. Cancer Imaging. 2005;**11**(5):S83-S88

[70] Alfuth R, Barnard CP. Family physicians and family therapists: Understanding the interdependent synergism. Contemporary Family Therapy. 2000;**22**(3):253-277

[71] Nancarrow SA, Borthwick AL. Dynamic professional boundaries in the healthcare force workforce. Sociology of Health & Illness. 2005;**27**(7):897-919

[72] Wilson SJ. The myth of objectivity: Is medicine moving towards a social constructivist paradigm? Family Practice. 2000;**17**(2):203-209

[73] Price P. Evidence-based laboratory medicine: Supporting decision-making. Clinical Chemistry. 2000;**46**(8):1041-1050

[74] Paul SA, Reddy MC. Understanding together: Sensemaking in collaborative information seeking. In: 2010 ACM Conference on Computer Supported Cooperative Work, CSCW 2010. 2010. pp. 321-330. DOI: 10.1145/1718918.1718976

[75] Everitt S. Clinical decision making in veterinary practice [doctoral thesis]. United Kingdom: University of Nottingham; 2011. Available from: http://etheses.nottingham.ac.uk/2051/ [Retrieved: 7 October 2012]

[76] Tatsioni A, Zarin DA, Aronson N, Samson DJ, Flamm CR, Schmid C, et al. Challenges in systematic reviews of diagnostic technologies. Annals of Internal Medicine. 2005;**142**(12, Part 2): 1048-1055

[77] Mol A, Elsman B. Detecting disease and designing treatment: Duplex and the diagnosis of diseased leg vessels. Sociology of Health and Illness. 1996;**18**(5):609-631

[78] Murphy FJ. The paradox of imaging technology: A review of the literature. Radiography. 2006;**12**(2):169-174

[79] Timmermans S, Berg M. The practice of medical technology. Sociology of Health & Illness. 2003;**25**(3):97-114

[80] Moreira T, Rapley T. Understanding the shaping, incorporation and coordination of health technologies through qualitative research. In: Bourgeault I, Dingwall R, De Vries R, editors. The SAGE Handbook of

Qualitative Methods in Health Research. 1st ed. London: Sage; 2010. pp. 658-672

[81] Reeves PJ, Decker S. Diagnostic radiography: A study in distancing. Radiography. 2012;**18**(2):78-83

Chapter 6

Dosimetry for Use in Preparedness and Response to Radiological and Nuclear Emergency

Vladimir Kutkov

Abstract

Lessons learned from responses to past radiological and nuclear emergencies have shown that more guidance is needed for assessing doses to those who were affected in emergency exposure situation. The chapter introduces system of dosimetric quantities for use in emergency preparedness and response to nuclear or radiological emergency, which includes RBE-weighted absorbed dose in tissue or organ for evaluation of the risk of severe deterministic effects, equivalent dose in tissue or organ for evaluation of the risk of stochastic effects, and effective dose for evaluation of the detriment due to undetectable stochastic effects. The chapter also provides internationally proved criteria for protection of individual in emergency exposure situation and framework of dose and risk assessment in an emergency. The special attention has been put on evaluation of available sources of dosimetric data needed for dose and risk assessment in an emergency.

Keywords: emergency, dosimetry, internal exposure, external exposure, deterministic effects, stochastic effects

1. Introduction

In event of nuclear or radiological emergency, uncontrolled exposure to ionizing radiation can cause fatal or threatening health effects. Dose received by individual determines the nature of such effects and their severity. To protect individual in emergency exposure situation, one needs to assess doses, which have been received, or would be received in emergency exposure situation, to be able to take an informed decision on protective and other response actions, including medical treatment of overexposed person as demanded in the IAEA General Safety Requirements (GSR) Part 7 [1]. The chapter presents a basis for dose assessment in emergency exposure situation and includes three parts:

1. Explanation of dosimetric quantities, such as basic, protection, and operational quantities, for use in emergency preparedness and response.

2. Criteria for assessment of doses received or expected to be received in emergency exposure situation.

3. Framework of estimation of protection quantities from monitoring results in event of nuclear or radiological emergency.

2. Dosimetric quantities

The International Commission on Radiation Units and Measurements (ICRU) in Report 30 [2] formulates a quantitative dosimetric concept in radiobiology as follows:

1. Biological effects of radiation are correlated with the energy absorbed by ionization and excitation in unit mass of tissue.

2. Biological effects of radiation are modified by microscopic spatial distribution of energy of radiation imparted to matter.

Dosimetric concept considers two-step assessment of a dose to characterize health consequences of exposure to ionizing radiation. On the first step, an energy of radiation absorbed in a tissue or organ has to be evaluated. On the next step, a human exposure has to be evaluated in term of protection quantities which could be used in models of developing radiation health effects to assess health consequences (risks) associated with irradiation. At this step, quality of radiation has to be considered taking into account its dependence on properties of radiation, properties of tissue or organ, and expected health effect.

2.1 Basic physical quantities

The basic dosimetric quantities include the particle fluence Φ_R, the kerma K_R, and the absorbed dose D_R of radiation R. Basic quantities characterize the field of radiation and its interaction with a medium where the human body could be present:

1. The fluence, Φ_R, is the quotient of dN_R by ds, where dN_R is the number of particles R incident on a sphere of cross-sectional area ds. Unit of fluence is m^{-2}.

2. The kerma, K_R, for ionizing uncharged particles R, is the quotient of dE_{tr} by dm, where dE_{tr} is the mean sum of the initial kinetic energies of all the charged particles liberated in a mass dm of a material by the uncharged particles incident on dm. Unit of kerma is J/kg. The special name for the unit of kerma is gray (Gy).

3. The absorbed dose, D_R, is the quotient of $d\bar{\varepsilon}$ by dm, where $d\bar{\varepsilon}$ is the mean energy imparted by ionizing radiation to matter of mass dm. Unit of absorbed dose is J/kg. The special name for the unit of absorbed dose is gray (Gy).

The ICRU provides exact definition of these quantities in [3].

2.2 Protection quantities

Protection quantities are the dosimetric quantities, which characterize the irradiation of the human. GSR Part 7 [1] states that in response to emergency, "consideration shall be given to actions to be taken to avoid or to minimize severe deterministic effects and to reduce the risk of stochastic effects. Deterministic effects shall be evaluated on the basis of relative biological effectiveness (RBE) weighted absorbed dose in tissue or organ. Stochastic effects in a tissue or organ shall be evaluated on the basis of equivalent dose in tissue or organ. The detriment associated with the occurrence of stochastic effects in individuals in an exposed population shall be evaluated on the basis of the effective dose" (para. 4.28 of GSR Part 7 [1]). Dosimetric quantities of RBE-weighted absorbed dose in tissue or organ T, AD_T; equivalent dose in tissue or organ T, H_T; and effective dose E are defined in GSG-2 [4] and GSR Part 3 [5] and recommended for protection purposes.

The overall theory on the development of biological health effects is not yet developed. The protection quantities are essentially used for evaluation of consequences of human exposure to ionizing radiation in terms of developing deterministic and stochastic health effects separately as listed in **Table 1**. **Figure 1** presents relationship between protection quantities and basic physical quantities.

Radiation protection quantities could not be measured directly. In particular irradiation conditions, they could be calculated by taking into account estimates of basic protection quantities (results of their measurement or calculation), geometry of irradiation, characteristics of irradiated person, etc. For radiation protection purposes, these quantities are defined for reference persons representing different groups of the public and workers in [6] and standard irradiation geometries for external exposure in [7].

Definitions of these quantities can be found in the GSR Part 3 [5], 2007 Recommendations of the International Commission on Radiological Protection (ICRP) [8], and EPR Publication [9].

The determination of RBE-weighted dose AD_T in a tissue or organ T involves the use of tissue-specific and radiation-specific factors $RBE_{T,R}$ as a multiplier of absorbed dose for radiation R, to reflect the relative biological effectiveness of the radiation in inducing severe deterministic effect in organ T at high doses:

$$AD_T = \sum_R D_{T,R} \times RBE_{T,R}, \quad (1)$$

where $D_{T,R}$ is the average absorbed dose in the tissue or organ T for radiation R.

Factor $RBE_{T,R}$ depends on the quality of radiation and macroscopic distribution of energy of radiation imparted to matter of affected organ or tissue [9]. The role of such heterogeneity is significant in the case of internal exposure [9, 10]. The RBE-weighted absorbed dose in tissue or organ is defined for use together with

Dosimetric quantity	Symbol	Unit	Purpose
Protection quantities			
RBE-weighted absorbed dose in tissue or organ T	AD_T	Gy	For evaluating deterministic effects induced as a result of exposure of an organ or tissue T
Equivalent dose in tissue or organ T	H_T	Sv	For evaluating stochastic effects induced as a result of exposure of an organ or tissue T
Effective dose	E	Sv	For evaluating detriment related to the occurrence of stochastic effects in an exposed population
Operational quantities			
Personal dose equivalent	$H_P(d)$	Sv	For monitoring external exposure of an individual
Ambient dose equivalent	$H^*(d)$	Sv	For monitoring a radiation field of the strongly penetrating radiation
Directional dose equivalent	$H'(d, \Omega)$	Sv	For monitoring a radiation field of the weakly penetrating radiation

Table 1.
Dosimetric quantities.

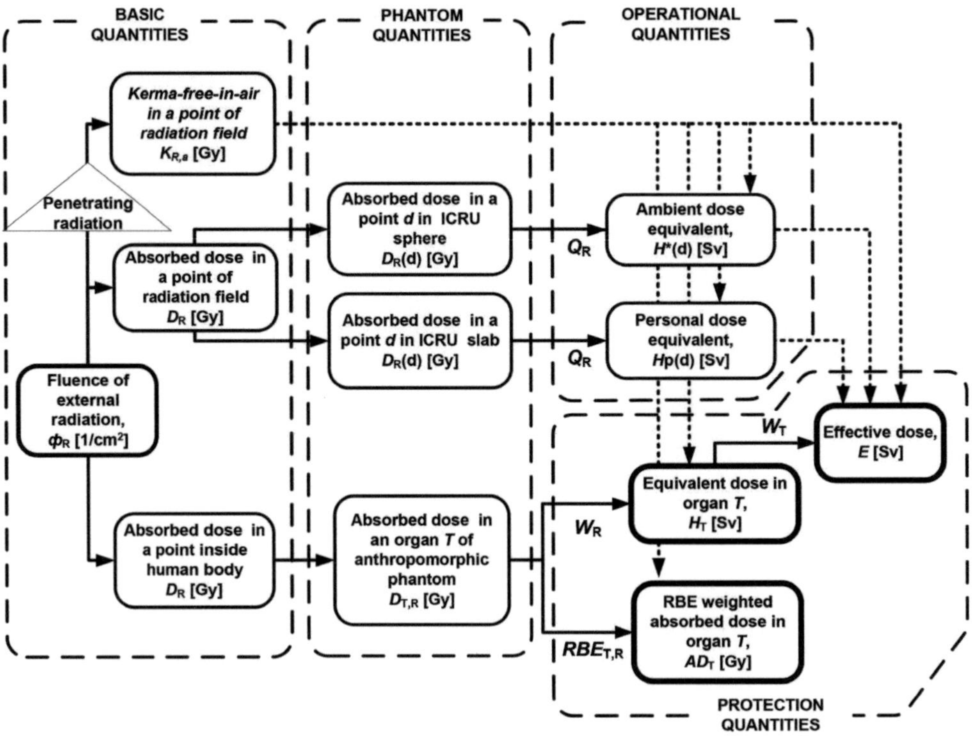

Figure 1.
Dosimetric quantities.

radiobiological model for assessing the risk of developing severe deterministic effects in critical organs and tissues at high doses [9–12].

The determination of equivalent dose H_T in a tissue or organ T involves the use of a radiation weighting factor w_R as a multiplier of absorbed dose for radiation R, to reflect the relative biological effectiveness of the radiation in inducing stochastic effects at low doses:

$$H_T = \sum_R D_{T,R} \times w_R. \tag{2}$$

The equivalent dose in tissue or organ is defined for use together with radiobiological model for assessing the risk of developing stochastic effects in organs and tissues at low doses of external [13] and internal [14] exposure.

The determination of effective dose E involves the use of a tissue weighting factor w_T as a multiplier of equivalent dose for tissue T, to account for the different sensitivities of different tissues or organs to the induction of stochastic effects and production of radiation detriment [8, 15]:

$$E = \sum_T H_T \times w_T = \sum_{T,R} D_{T,R} \times w_R \times w_T. \tag{3}$$

In accordance with the ICRP, the radiation detriment is the total harm to health experienced by an exposed group and its descendants as a result of the group's exposure to a radiation source. Detriment is a multidimensional concept. Its principal components are the stochastic quantities: probability of attributable fatal cancer, weighted probability of attributable nonfatal cancer, weighted probability of severe heritable effects, and length of life lost if the harm occurs [8]. The 2007

Recommendations of the ICRP state in para. (153) that the main and primary uses of effective dose in radiological protection for both occupational workers and the general public are:

- prospective dose assessment for planning and optimization of protection; and
- retrospective dose assessment for demonstrating compliance with dose limits, or for comparing with dose constraints or reference levels.

In this regard, organ or tissue equivalent doses, not effective doses, are required for assessing the probability of cancer induction in exposed individuals.

The recommended values of $RBE_{T,R}$, w_R and w_T are based on a review of published biological and epidemiological studies and are given in the definitions of protection quantities in GSR Part 3 [5].

2.3 Operational quantities

Since radiation protection quantities cannot be measured directly, the ICRU introduced operational quantities for practical use in radiation protection where exposure due to external sources is concerned. Definitions of these quantities can be found in [16–20] and in GSR Part 3 [5]. The operational quantities provide an estimate of effective or equivalent dose in tissue or organ in such a way that avoids overestimation in most radiation fields encountered in practice [7]. Radiation quality factor Q(L) is used in calculating the operational dose equivalent quantities used in monitoring [21]. The quality factor characterizes the biological effectiveness of the radiation type, based on the ionization density along the tracks of charged particles in tissue. Q is defined as a function of the unrestricted linear energy transfer, L_m (often denoted as L or linear energy transfer, LET), of charged particles in water. Detailed evaluation of the relationship between the physical, protection, and operational quantities was conducted by a joint task group of the ICRP and ICRU [7].

Strongly penetrating radiation and weakly penetrating radiation are considered in radiation dosimetry and are differentiated as follows. For most practical purposes, it may be assumed that strongly penetrating radiation includes photons of energy above about 12 keV, electrons of energy more than about 2 MeV, and neutrons. It may be also assumed that weakly penetrating radiation includes photons of energy below about 12 keV, electrons of energy less than about 2 MeV, and massive charged particles as protons and alpha particles [5].

The operational quantity for individual monitoring is the personal dose equivalent $H_P(d)$. Any statement of personal dose equivalent has to include a specification of the reference depth d. In order to simplify the notation, d is assumed to be expressed in millimeters.

For strongly penetrating radiation, the reference depth for controlling the radiation detriment in planned exposure situation is 10 mm. Personal dose equivalent $H_P(10)$ provides a conservative estimate of effective dose to adult for strongly penetrating radiation.

For weakly penetrating radiation, the reference depth for controlling stochastic effects due to irradiation of the basal membrane of the skin is 0.07 mm, and the deterministic effect in lens of the eye is 3 mm [15, 22, 23].

For monitoring of the lens of the eye, a depth of 3 mm is recommended by the International Commission on Radiation Units and Measurements [3], so the operational quantity to be used is $H_P(3)$. In practice, however, the use of $H_P(3)$ has not yet been implemented for routine individual monitoring. In specific cases, when actual workplace radiation fields are known, monitoring of the eye through dosimeters calibrated for $H_P(0.07)$ could be acceptable. In [23], it is stated that $H_P(0.07)$ can be considered a good operational quantity for the lens of the eye for exposures to fields for which most of the dose is due to photons, including X radiation. In such cases, it has to be borne in mind that the uncertainty associated with the estimation of equivalent dose will be higher.

The operational quantities recommended for workplace monitoring are defined in a phantom known as the ICRU sphere [18]. This is a sphere of 30 cm diameter made of tissue equivalent material with a density of 1 g/cm^3 and an elemental composition (by mass) of 76.2% oxygen, 11.1% carbon, 10.1% hydrogen, and 2.6% nitrogen.

The two quantities recommended by the International Commission on Radiation Units and Measurements for workplace monitoring [3] are the ambient dose equivalent $H^*(d)$ and the directional dose equivalent $H'(d, \Omega)$.

The ambient dose equivalent $H^*(d)$ at a point in a radiation field is the dose equivalent that would be produced by the corresponding aligned and expanded field in the ICRU sphere, at a depth d on the radius opposing the direction of the aligned field.

The expanded field (see **Figure 2(b)**) is one in which the fluence and its angular and energy distribution are the same throughout the volume of interest as in the actual field at the point of reference (see **Figure 2(a)**). In the expanded and aligned field (see **Figure 2(c)**), the fluence and its energy distribution are the same as in the expanded field, but the fluence is unidirectional [12].

Any statement of ambient dose equivalent has to include a specification of the reference depth d. For strongly penetrating radiation, the recommended depth is 10 mm. When measuring $H^*(10)$, the radiation fields have to be uniformed over the sensitive volume of the instrument, and the instrument has to have an isotropic response.

The directional dose equivalent $H'(d, \Omega)$ at a point in a radiation field is the dose equivalent that would be produced by the corresponding expanded field in the ICRU sphere, at a depth d on a radius in a specified direction Ω. Any statement of directional dose equivalent has to include a specification of the reference depth d and the direction Ω of the radiation. For strongly penetrating radiation and weakly penetrating radiation, the recommended depths are 10 mm and 0.07 mm, respectively.

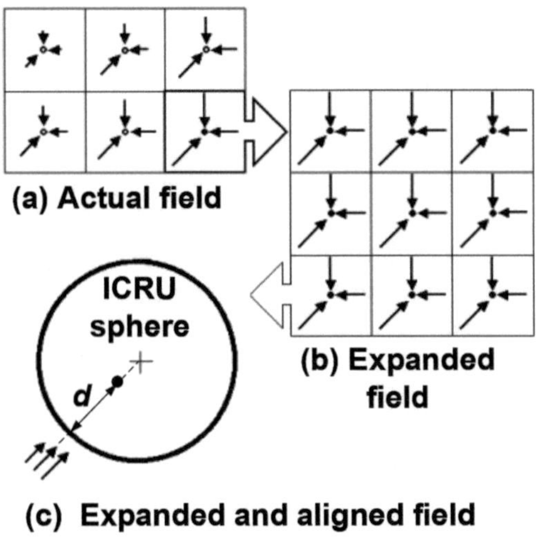

Figure 2.
Actual, expanded, and aligned radiation field.

If the field is unidirectional, the direction Ω is specified as the angle between the radius opposing the incident field and the specified radius. When the specified radius is parallel to the radiation field (i.e., when Ω = 0°) the quantity $H'(d, 0)$ is equal to $H^*(d)$. When measuring $H'(d, \Omega)$, the radiation fields have to be uniformed over the dimensions of the instrument, and the instrument has to have the appropriate directional response.

3. Generic dosimetric criteria

When determining the possible health hazard, three dosimetric quantities must be considered as shown in **Table 1**:

1. The RBE-weighted dose AD_T, which is used to evaluate the risk of severe deterministic effects

2. The equivalent dose H_T, which is used to evaluate the risk of stochastic effects

3. The effective dose E, which is used to evaluate the radiogenic detriment for purposes of radiation protection

Protection quantities are applicable in certain dose ranges. **Figure 3** provides an example of the applicable ranges for dosimetric quantities characterizing external penetrating radiation to evaluate risk of severe deterministic effects and stochastic effects, i.e., observable increase in the incidence of radiation-induced cancers. The ranges are not exactly defined because of competition of effects. For instance, both the risks of radiogenic cancers and severe deterministic effects have to be evaluated when whole-body absorbed dose is around 1 Gy. Below 100 mGy may not have any severe deterministic effects or an observable increase in the incidence of cancer, even in a very large exposed group. An increase in the cancer incidence rate due to radiation-induced cases is uncertain and will not be detectable [10, 12, 24].

The system of protective actions and other response actions in an emergency includes numerical values of generic criteria as well as of the corresponding operational criteria that form the basis for decision-making in an emergency.

Table 2 presents levels of RBE-weighted absorbed dose in critical organs and tissues which if exceeded will give rise to severe deterministic effects in 5% of those who are exposed.

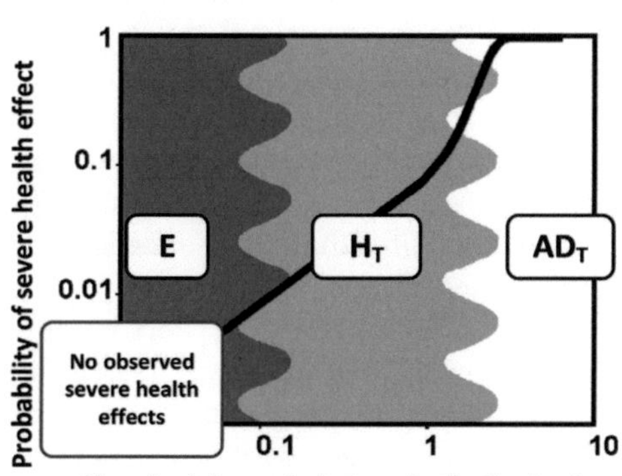

Figure 3.
Probability of severe health effects as a function of absorbed dose for external penetrating radiation.

Quantity	Level	Comments
Acute external exposure (for less than 10 h)		
$AD_{Red\ marrow}$	1 Gy	$AD_{Red\ marrow}$ represents the RBE-weighted absorbed dose in internal tissues or organs from exposure in a uniform field of strongly penetrating radiation [1, 4]
$AD_{Fetus/Embryo}$	0.1 Gy	No
$AD_{Soft\ tissue}$	25 Gy	Dose delivered to 100 cm^2 at a depth of 0.5 cm under the body surface in soft tissue [1, 4]
$AD_{Skn\ derma}$	10 Gy	Dose delivered to the 100 cm^2 dermis (skin structures at a depth of 40 mg/cm^2 (or 0.4 mm) below the body surface) [1, 4]
Acute intake of radioactive substance		
$AD_{Red\ marrow}(\Delta)$	0.2 Gy	For intake of radionuclides with $Z > 90$; $\Delta = 30$ d [1, 4]
	2 Gy	For intake of radionuclides with $Z \leq 89$; $\Delta = 30$ d [1, 4]
$AD_{Thyroid}(\Delta)$	2 Gy	For intake of thyroid-seeking radionuclides; $\Delta = 30$ d [1, 4]
$AD_{Lung}(\Delta)$	30 Gy	For the purposes of these generic criteria, "lung" means the alveolar-interstitial region of the respiratory tract; $\Delta = 30$ d [1, 4]
$AD_{Colon}(\Delta)$	20 Gy	For the purposes of these generic criteria, "colon" presents upper and lower large intestine: $\Delta = 30$ d [1, 4]
$AD_{Fetus/Embryo}\left(\tilde{\Delta}\right)$	0. 1 Gy	For this particular case, $\tilde{\Delta}$ means the period of in utero development [1, 4]

Table 2.
Criteria for assessing the high doses from external and internal exposure.

Quantity	Level	Comments
E	100 mSv in 1 year	No
$H_{Fetus/Embryo}$	100 mSv in 1 year	No
$H_{Thyroid}$	100 mSv in 1 year	For intake of thyroid-seeking radionuclides

Table 3.
Criteria for assessing the intermediate doses from external and internal exposure.

Table 3 provides the dosimetric criteria used to define the radiological health hazard level used for taking a decision on implementation of protective and other response actions in emergency exposure situation. These criteria are based on **Table 3** of [4], which provides the international generic criteria below which the risk of cancers and other health effects is too low to justify taking any protective or other response actions, such as a medical screening. The criteria were established for exposures at high-dose rates. For the lower-dose rates that will occur off the site following a release resulting from a severe emergency at a light-water reactor or its spent fuel pool, a comparable level of radiation-induced cancer risk would probably occur at a dose two or more times higher [8].

For restricting the exposure of emergency workers having assigned tasks in an emergency response, **Table 4** provides guidance values in terms of personal dose equivalent $H_P(10)$ from external exposure to strongly penetrating radiation. The values for $H_P(10)$ in **Table 4** assume that every effort has been made for protection against external exposure to weakly penetrating radiation and against exposure due to intakes or skin contamination (see para. 5.53 of GSR Part 7 [1]). The values of

Task	Quantity and guidance level		
	$H_P(\mathbf{10})$	E	AD_T
Lifesaving actions [(a)] Actions to prevent severe deterministic effects and actions to prevent the development of catastrophic conditions that could significantly affect people and the environment	500 mSv	500 mSv	50% from **Table 2**
Actions to avert a large collective dose	100 mSv	100 mSv	10% from **Table 2**

*Note: [(a)] This value may be exceeded—with due consideration of the generic criteria in **Table 2**—under circumstances in which the expected benefits to others clearly outweigh the emergency worker's own health risks and the emergency worker volunteers to take the action and understands and accepts these health risks [1].*

Table 4.
Criteria for assessing individual doses in emergency workers.

$H_P(10)$ in **Table 4** are used for planning and operational monitoring of the work's exposure.

To assure protection of the emergency workers, the total effective dose and the RBE-weighted absorbed dose in a tissue or organ via all exposure pathways need to be estimated as early as possible in a nuclear or radiological emergency.

4. Estimation of protection quantities

Protection quantities of AD_T, H_T and E characterize the exposure of an individual for purposes of implementation of protective actions and other actions to protect him or her in emergency exposure situation. The practical goal of radiation monitoring in emergency exposure situation is to provide the information required for estimation of protective quantities.

4.1 Estimation of protection quantities characterizing an external exposure

Properties of external radiation entering the body of individual R, basic physical characteristics of radiation field, such as particle fluence Φ_R, the kerma K_R, and geometry of irradiation G, provide a basis for estimation of the protection quantities from external exposure. For idealized irradiation geometries, the ICRP in [7] presents the relationship between basic physical quantities and protection quantities for monoenergetic photons, electrons and positrons, neutrons, protons, muons, pions, and helium ions in six considered irradiation geometries: AP, anteroposterior; PA, posteroanterior; LLAT, left lateral; RLAT, right lateral; ROT, rotational; and ISO, isotropic.

As presented in **Figure 4**, the major characteristic of a field of radiation is a fluence of radiation Φ_R. For monoenergetic radiation, an ambient dose equivalent and air kerma are proportional to fluence:

$$\Phi_R = K_{a,R}/k_{a,R} = H_R^*(d)/h_R^*(d). \tag{4}$$

For non-monoenergetic radiation

$$K_a = \sum_R \int \Phi_R(\varepsilon_R) \times k_a(\varepsilon_R) d\varepsilon_R, \tag{5}$$

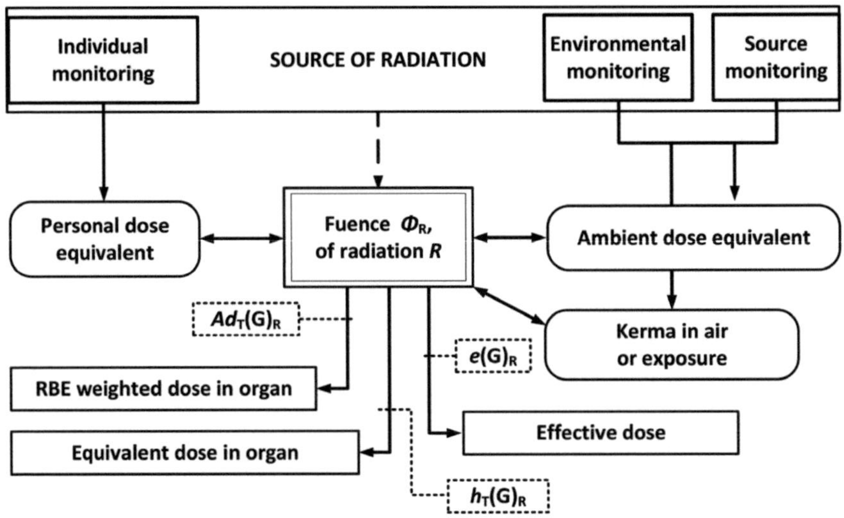

Figure 4.
General scheme for assessment of individual external dose from monitoring.

$$H^*(10) = \sum_R \int \Phi_R(\varepsilon_R) \times h^*(10, \varepsilon_R) d\varepsilon_R, \tag{6}$$

where $\Phi_R(\varepsilon_R) d\varepsilon_R$ is the fluence of radiation R with energy between ε_R and $\varepsilon_R + d\varepsilon_R$, $k_a(\varepsilon_R)$ is the dose coefficient of air kerma of radiation R with energy ε_R, and $h^*(10, \varepsilon_R)$ is the dose coefficient of ambient dose equivalent of radiation R with energy ε_R.

For irradiation geometry G, one has

$$H_P(10)_G = \sum_R \int \Phi_R(\varepsilon_R) \times h_P(10, \varepsilon_R)_G d\varepsilon_R, \tag{7}$$

$$AD_{T,G} = \sum_R \int \Phi_R(\varepsilon_R) \times Ad_T(10, \varepsilon_R)_G d\varepsilon_R, \tag{8}$$

$$H_{T,G} = \sum_R \int \Phi_R(\varepsilon_R) \times h_T(\varepsilon_R)_G d\varepsilon_R, \tag{9}$$

$$E_G = \sum_R \int \Phi_R(\varepsilon_R) \times e(\varepsilon_R)_G d\varepsilon_R, \tag{10}$$

where $k_a(\varepsilon_R)$ is the dose coefficient of air kerma of radiation R with energy ε_R, $h_P(10, \varepsilon_R)_G$ is the dose coefficient of personal dose equivalent of radiation R with energy ε_R and irradiation geometry G, $Ad_T(\varepsilon_R)_G$ is the dose coefficient of RBE-weighted dose in organ T of radiation R with energy ε_R and irradiation geometry G, $h_T(\varepsilon_R)_G$ is the dose coefficient of equivalent dose in organ T of radiation R with energy ε_R and irradiation geometry G, and $e(\varepsilon_R)_G$ is the dose coefficient of effective dose of radiation R with energy ε_R and irradiation geometry G.

For photon radiation, values of $Ad_T(\varepsilon_R)_G$ and $h_T(\varepsilon_R)_G$ are numerically equal when the same organ or tissue and irradiation geometry are considered.

The internet resource [25] provides tools for evaluation of absorbed dose in different organs from point or volumetric radioactive sources inside or outside the human body.

The publication [26] provides dose coefficients for exposure to bulk sources that are ground, water body and cloud containing radioactive material.

The publication [27] presents a compendium of neutron spectra, which could be used for estimation of protection quantities in accordance with Eqs. (7)–(10). The estimates show that for fission neutrons scattered from the concrete walls of the facility, soil, or from the air surrounding the facility (skyshine), the value of $Ad_T(\varepsilon_R)_G$ is numerically equal to 1/5 of $h_T(\varepsilon_R)_G$ when the same organ or tissue and irradiation geometry are considered. This linkage provide the possibility to use results of operational or routine monitoring of doses in workers for estimation of $AD_{Red\ marrow}$ in emergency exposure situation as required by GSR Part 7 and presented in **Table 4**.

As presented in [7], dose coefficients mentioned above are proportional to the photon energy in range of (0.1–6) MeV. Therefore, for the same irradiation geometry

$$\frac{H_{T,G} \times \varepsilon_j}{h_T(\varepsilon_j)_G} \cong \frac{H_P(10)_G \times \varepsilon_j}{h_P(10, \varepsilon_j)_G} \cong \frac{H^*(10) \times \varepsilon_j}{h^*(10, \varepsilon_j)} \cong \frac{K_a \times \varepsilon_j}{k_a(\varepsilon_j)} \ldots, \tag{11}$$

where ε_j is any photon energy from range of (0.1–6) MeV.

For mentioned range of photon' energy, the exposure is proportional to the kerma free-in-air (air kerma) or exposure. Thus, in the same point of field of photon radiation, exposure or kerma in air for an exposure of 100 R is 0.876 Gy [28]. This linkage provides the possibility to use old devices such as exposure meters (R-meters) in environmental monitoring for estimation of air kerma and protection quantities as given in Eq. (11).

The protection quantities AD_T, H_T and E received from exposure due to external sources can be also estimated from the operational quantities by using the following equations:

$$E \cong H_P(10). \tag{12}$$

$$H_{Skin} \cong H_P(0.07). \tag{13}$$

$$H_{Lense\ of\ eye} \cong H_P(0.07). \tag{14}$$

For strongly penetrating radiation, the critical organ for controlling the development of the severe deterministic effects in individual is the red marrow [1, 4]. The ICRU did not recommend depth for controlling the dose in the red marrow. For practical reasons, monitoring of the red marrow through dosimeters calibrated for $H_P(10)$ could be acceptable. The RBE-weighted absorbed dose in the red marrow received from exposure due to external sources can be estimated from the operational quantities by using the following equation:

$$AD_{Red\ Marrow} \cong H_P(10). \tag{15}$$

For weakly penetrating radiation and in emergency exposure situation, the reference depth for controlling severe deterministic effects due to irradiation of the derma of the skin is 0.4 mm and for controlling the severe deterministic effects due to irradiation of shallow soft tissue is 5 mm [1, 4, 9, 10]. The ICRU did not recommend operational quantity for controlling the dose in the skin derma or shallow soft tissue in emergency exposure situation. For practical reasons, monitoring of the skin derma and shallow soft tissue through dosimeters calibrated for $H_P(0.07)$ could be acceptable. The RBE-weighted absorbed dose in the skin derma or in shallow soft tissue received from exposure due to external sources can be conservatively estimated from the operational quantities by using the following equations:

$$AD_{Skin\ derma} \cong (0.07). \tag{16}$$

$$AD_{Soft\ tissue} \cong H_P(3). \tag{17}$$

Based on the foregoing, individual monitoring in planned, emergency and existing exposure situations through individual dosimeters calibrated for $H_P(10)$ and $H_P(0.07)$ could be acceptable.

4.2 Estimation of protection quantities characterizing an internal exposure

Internal doses cannot be measured directly; they can only be calculated from intake of radioactive substance through particular route such as respiratory system in the case of inhalation, gastrointestinal tract in the case of ingestion or wound, and undamaged skin in the case of contamination. The individual intake also could not be directly measured and can only be inferred from individual measurements of other quantities, such as measurements of activity in the body or in excretion samples or activity concentration in foodstuff or the environment. In circumstances where individual monitoring is inappropriate, inadequate, or not feasible, the occupational exposure of workers may be assessed on the basis of workplace monitoring and other relevant information such as location and durations of exposure. Individual measurements include measurements made by both direct and indirect methods. Methods for the measurement of activity content in the body, such as whole-body, lung, or thyroid counting, are examples of direct methods. Measurements of activity in collected biological samples or measurements made using personal air sampling are examples of indirect methods.

The conceptual framework for the assessment of doses from individual or environmental measurements is illustrated in **Figure 5**.

Chemical and physical properties of radioactive substance S containing radionuclide, the route of intake into the human body P, and value of intake $I_{S,P}$ composite a base for estimation of the protection quantities from internal exposure.

The measurable characteristics of internal exposure are quantities of body content or excretion rate of radionuclide M and concentration of radioactive substance in the environmental media $C(t)_S$.

According to the scheme in **Figure 5**, the value of the intake of radioactive substance S could be obtained by multiplying the integrated over time the measured concentration of radioactive substance in the environmental media $C(t)_S$ by the appropriate value of $v_P(g)$:

$$I(g)_{S,P} = v_P(g) \times \int_0^t C(x)_S dx, \tag{18}$$

where $v_P(g)$ is the consumption rate of an individual of age group g through route P and t is the period of consumption.

According to the scheme in **Figure 5**, the value of the intake of radioactive substance S is obtained by dividing the measured body content or excretion rate M by the appropriate value of $m(t,g)_{S,P}$:

$$I(g)_{S,P} = M/m(t,g)_{S,P}, \tag{19}$$

where $m(t,g)_{S,P}$ is the fraction of an intake that remains in the body (for direct methods) or that is being excreted from the body (for indirect methods) at time t after the intake. This fraction depends on the radionuclide, its chemical and physical form in substance S, the route of intake P, the age group of age g, and the time t [29, 30].

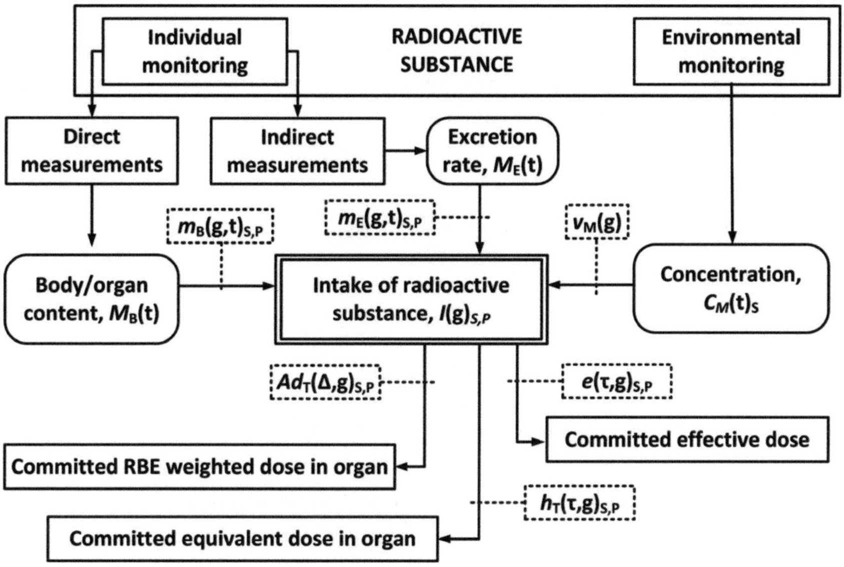

Figure 5.
General scheme for assessment of individual internal dose from monitoring.

Special attention has to be paid to the interpretation of bioassay measurements after the use of means for blocking the uptake of radionuclides or for enhancing their excretion, such as the administration of diuretics, laxatives, or blocking or chelating agents, as well as after the removal of contamination and/or surgical intervention at a wound site. These techniques influence and modify the biokinetic behavior of the incorporated radionuclides, thus invalidating the use of the standardized modeling approach for estimating intake and dose from the bioassay measurements.

In such cases, alternative approaches have to be employed, such as discarding data on excretion for excretion samples collected during the period in which excretion rates may be assumed to have been influenced by the treatment or modifying the standard models in order to take into account the effect of the treatment. Examples of analyses performed after the administration of the chelating agent Ca-DTPA (a calcium salt of diethylenetriaminepentaacetic acid) in cases of accidental intakes of actinides can be found in [31–33]. Bioassay measurements for dose assessment purposes are performed after a certain time period, posttreatment with Ca-DTPA, until the excretion of the radionuclide stabilizes in urine samples. Influence of iodine thyroid blocking on biokinetic of iodine in human body is discussed in [34]. Influence of administration of Prussian blue $\{Fe_4[Fe(CN)_6]_3\}$ on biokinetic of Cs is discussed in [35].

The specific protection quantities of committed RBE-weighted organ dose $AD_T(\Delta,g)$, committed equivalent organ dose $H_T(\tau,g)$, and committed effective dose $E(\tau,g)$ are used for evaluation of dose in an internal dosimetry:

$$AD_T(\Delta,g) \cong \sum_{S,P} \left\{ I(g)_{S,P} \times Ad_T(\Delta,g)_{S,P} \right\}, \tag{20}$$

$$H_T(\tau,g) = \sum_{S,P} \left\{ I(g)_{S,P} \times h_T(\tau,g)_{S,P} \right\}, \tag{21}$$

$$E(\tau,g) = \sum_{S,P} \left\{ I(g)_{S,P} \times e(\tau,g)_{S,P} \right\}, \tag{22}$$

where $Ad_T(\Delta,g)_{S,P}$ and $h_T(\tau,g)_{S,P}$ are respectively the committed RBE-weighted or equivalent dose in the organ or tissue T due to intake of 1 Bq of radionuclide substance S through pathway P, by the group of age g. These dose factors are the time integrals of the relevant dose rates:

$$Ad_T(\Delta,g)_{S,P} = \int_0^{\Delta} \dot{Ad}_T(t,g)_{S,P}\,dt, \tag{23}$$

$$h_T(\tau,g)_{S,P} = \int_0^{\tau} \dot{h}_T(\mathrm{t},g)_{S,P}\,dt, \tag{24}$$

$$e(\tau,g)_{S,P} = \sum_T \left\{h_T(\tau,g)_{S,P} \times w_T\right\}, \tag{25}$$

where $\dot{Ad}_T(t,g)_{S,P}$ is the RBE-weighted dose rate in organ or tissue T due to intake of 1 Bq of radionuclide substance S through pathway P and $\dot{h}_T(\mathrm{t},g)_{S,P}$ is the equivalent dose rate in organ or tissue T due to intake of 1 Bq of radionuclide substance S through pathway P. For estimation of committed RBE-weighted dose, Δ is taken to be 30 days. The τ is the integration time elapsed after an intake of radioactive substances. When τ is not specified, it will be taken to be 50 years for adults and the time to age 70 years for intakes by children and persons younger than 20.

Various biokinetic models for calculating the values of $m(t,g)_{S,P}$, $Ad_T(\Delta,g)_{S,P}$ and $h_T(\tau,g)_{S,P}$ have been developed.

Values of $m(t,g)_{S,P}$ at selected times for a subset of radionuclides have been reported by the ICRP in graphical and tabular form [36, 37]. A compilation of $m(t,g)_{S,P}$ by workers in emergency and planned exposure situation is presented in various publications of the IAEA [29, 38]. The Internet resources containing the values of $m(t,g)_{S,P}$ different radioactive substances are presented in [39].

Special attention has to be paid to the patterns of inhalation intake in nuclear emergency. Chemical form of radionuclides in aerosol particles and their behavior in the human respiratory tract could be very specific [40, 41] and significantly different from these observed in planned exposure situation and presented in Table II.2C of GSR Part 3 [5].

A compilation of dose coefficients $Ad_T(\Delta,g)_{S,P}$ for intakes of radionuclides by adults is presented in publications of the IAEA [9, 29].

A compilation of dose coefficients $e(\tau,g)_{S,P}$ for ingestion and inhalation intakes of radionuclides by workers is presented in ICRP Publication 119 [42]. A compilation of $h_T(\tau,g)_{S,P}$ and $e(\tau,g)_{S,P}$ for workers and all age groups of members of the public can also be found in the ICRP database in [43]. Data for fetus or embryo due to inhalation or ingestion of radioactive substance by mother are presented in [44, 45] and for infant in [46, 47]. These dose coefficients are based on the calculation methods and parameters given in ICRP Publication 60 [15]. The current published values of $e(\tau,g)_{S,P}$ will be superseded in due course by new values [6, 36, 42] based on updated biokinetic models and on the methods of calculation and the parameters given in ICRP Publication 103 [8]. A compilation of $m(t,g)_{S,\,Wond}$, $h_T(\tau,g)_{S,\,Wound}$ and $e(\tau,g)_{S,\,Wound}$ for penetration of radioactive substance through wound is presented in [48, 49].

4.3 Estimation of protection quantities characterizing a total exposure

The protection quantities E and H_T relate to the sum of the effective doses or equivalent doses, respectively, received from exposure due to external sources within a given time period and the committed effective doses or committed equivalent doses, respectively, from exposure due to intakes of radionuclides occurring within the same time period.

The total equivalent dose in organ or tissue T received or committed during a given time period in individuals from age group g can be estimated from the operational quantities by using the following equation:

$$H_T(g) \cong H_P(10) + H_T(\tau,g), \tag{26}$$

where $H_P(10)$ is the personal dose equivalent from external exposure and $H_T(\tau,g)$ is assessed by Eq. (21).

The total effective dose E received or committed during a given time period in individuals from age group g can be estimated from the operational quantities by using the following equation:

$$E(g) \cong H_P(10) + E(\tau,g), \tag{27}$$

where $E(\tau,g)$ is assessed by Eq. (22).

In the calculation of the committed dose from specific radionuclides, allowance may need to be made for the characteristics of the material taken into the body through inhalation, ingestion, and contaminated wound.

5. Conclusions

To characterize the emergency exposure situation in event of nuclear or radiological emergency, doses to the members of the public, workers, emergency workers, as well as patients and helpers, if applicable, have to be derived from source monitoring, environmental monitoring, or individual monitoring, or from a combination of these. Result of dose assessment needs to be expressed in terms of protection quantities defined in the IAEA General Safety Requirements (GSR) Part 7 [1] and GSR Part 3 [5]. Dose assessment has to be based on the best available monitoring data and has to be promptly updated if any new information relevant for dose assessment becomes available.

The dose assessment has to be as realistic as possible, and in any case, doses for situations in which persons might be in danger of being harmed are not to be underestimated. Overestimation also has to be avoided because there are risks associated with protective actions.

Author details

Vladimir Kutkov
National Research Centre, Kurchatov Institute, Moscow, Russia

*Address all correspondence to: v.kutkov@yandex.ru

References

[1] Preparedness and response for a nuclear or radiological emergency. General Safety Requirements Part 7. Safety Standard Series No. GSR-Part 7. Vienna: IAEA; 2015. p. 136

[2] Quantitative Concepts and Dosimetry in Radiobiology. ICRU Report 30. Vol. 75. Bethesda, MD: ICRU; 1979

[3] Fundamental quantities and units for ionizing radiation. ICRU Report 85a-Revised. Journal of the ICRU. 2011; **11**(1):35. Oxford: Oxford University Press

[4] Criteria for use in preparedness and response for a nuclear or radiological emergency. IAEA Safety Standards Series No. GSG-2. Vienna: IAEA; 2011. p. 120

[5] Radiation protection and safety of radiation sources: International basic safety standards. General Safety Requirements Part 3. Safety Standard Series No. GSR-Part 3. Vienna: IAEA; 2014. p. 471

[6] ICRP Publication 130. Occupational intakes of radionuclides: Part 1. Annals of the ICRP. 2015;**44**(2):188. Elsevier

[7] ICRP Publication 116. Conversion coefficients for radiological protection quantities for external radiation exposures. Annals of the ICRP. 2010;**40** (2-5):257. Elsevier

[8] ICRP Publication 103. The 2007 recommendations of the international commission on radiological protection. Annals of the ICRP. 2007;**37**(2-4):330. Elsevier

[9] Dangerous Quantities of Radioactive Material, Emergency. Preparedness and Response Series EPR D-VALUES. Vienna: IAEA; 2006. p. 154

[10] Extended Framework of Emergency Response Criteria: Interim Report for Comments. IAEA-TECDOC Series 1432. Vienna: IAEA; 2005. p. 108

[11] Kutkov V, Buglova E, McKenna T. Severe deterministic effects of external exposure and intake of radioactive material: Basis for emergency response criteria. Journal of Radiological Protection. 2011;**31**:237-253

[12] Kutkov V.A, Tkachenko V.V, Romantsov V.P. Radiation Safety of Personnel of Nuclear Power Plants. Edited by V.A. Kutkov. Moscow—Obninsk: Atomtekhenergo—IATE; 2003. 344 p. (In Russian)

[13] Report of the NCI-CDC Working Group to Revise the 1985 NIH Radioepidemiological Tables. U.S. Department of Health and Human Services; 2003. p. 128

[14] Cancer Risk Coefficients for Environmental Exposure to Radionuclides. Federal Guidance Report No.13. Washington DC: U.S. Environmental Protection Agency; 1999. p. 335

[15] ICRP Publication 60. 1990 Recommendations of the International Commission on Radiological Protection. Annals of the ICRP. 1991;**21**(1-3):199. Oxford: Pergamon Press

[16] Determination of Dose Equivalents from External Radiation Sources. ICRU Report 39. Bethesda, MD: ICRU; 1985. p. 14

[17] Determination of Dose Equivalents Resulting from External Radiation Sources Part 2. ICRU Report 43. Bethesda, MD: ICRU; 1988. p. 56

[18] Measurement of Dose Equivalents from External Photon and Electron

Radiations. ICRU Report 47. Bethesda, MD: ICRU; 1992. p. 45

[19] Determination of operational dose equivalent quantities for neutrons. ICRU Report 66. Journal of the ICRU 1 (3). Ashford, Kent: Nuclear Technology Publishing; 2001. 95 p

[20] Quantities and Units in Radiation Protection Dosimetry. ICRU Report 51. Bethesda, MD: ICRU; 1993. p. 19

[21] The Quality Factor in Radiation Protection. ICRU Report 40. Bethesda, MD: ICRU; 1986. p. 32

[22] Guidance on Radiation Dose Limits for the Lens of the Eye. NCRP Comments 26. Bethesda, MD: U.S. NCRP; 2016. p. 145

[23] Implications for Occupational Radiation Protection of the New Dose Limit for the Lens of the Eye, IAEA TECDOC Series 1731. Vienna: IAEA; 2013. p. 48

[24] Biological effects at low radiation doses. Sources and Effects of Ionizing Radiation. In: UNSCEAR 2000 Report to the General Assembly (with scientific annexes). Vol. II. Scientific Annex G. New York: United Nations; 2000. pp. 73-176

[25] Visual Monte Carlo. Software for radiation protection and internal dosimetry. Available from: http://www.VMCsoftware.com [Accessed: 01-12-2018]

[26] External Exposure to radionuclides in air, water and soil. Federal Guidance Report No. 15. Washington DC: U.S. Environmental Protection Agency; 2018. p. 342

[27] Compendium of Neutron Spectra and Detector Responses for Radiation Protection Purposes. Supplement to Technical Reports Series No. 318. Technical Report Series No. 403. Vienna: IAEA; 2001. p. 276

[28] Podgoršak EB. Radiation Physics for Medical Physicists. 3rd ed. Switzerland: Springer; 2016. p. 955

[29] Generic procedures for medical response during a nuclear or radiological emergency, Emergency Preparedness and Response Series EPR-MEDICAL. Vienna: IAEA; 2005. p. 296

[30] Occupational Radiation Protection. Safety Standards Series No. GSG-7. Vienna: IAEA; 2018. p. 360

[31] Bailey BR, Eckerman KF, Townsend LW. An analysis of a puncture wound case with medical intervention. Radiation Protection Dosimetry. 2003; **105**:509-512

[32] Fritsch P, Grappin L, Guillermin AM, et al. Modelling of bioassay data from a Pu wound treated by repeated DTPA perfusions: Biokinetics and dosimetric approaches. Radiation Protection Dosimetry. 2007;**127**:120-124

[33] James AC, Sasser LB, Stuit DB, et al. USTUR whole body case 0269: Demonstrating effectiveness of I.V. Ca-DTPA for Pu. Radiation Protection Dosimetry. 2007;**127**:449-455

[34] Zanzonico PB, Becker DV. Effects of time of administration and dietary iodine levels on potassium iodide (KI) blockade of thyroid irradiation by ^{131}I from radioactive fallout. Health Physics. 2000;**78**:660-667

[35] Melo DR, Lipsztein JL, Oliveira CAN, Bertelli L. ^{137}Cs internal contamination involving a Brazilian accident, and the efficacy of Prussian Blue treatment. Health Physics. 1994; **66**:645-652

[36] ICRP Publication 134. Occupational intakes of radionuclides: Part 2. Annals of the ICRP. 2016;**45**(3–4):349

[37] ICRP Publication 137. Occupational intakes of radionuclides: Part 3. Annals of the ICRP. 2017;**46**(3–4):486

[38] Methods for assessing occupational radiation doses due to intakes of radionuclides. Safety Reports Series No. 37. Vienna: IAEA; 2004. p. 124

[39] Graphs of Predicted Monitoring Data. Chiba: National Institute of Radiological Sciences. Available from: http://www.NIRS.qst.go.jp/db/anzendb/RPD/gpmd.php [Accessed: 01-12-2018]

[40] Kutkov VA, Arefieva ZS, Muraviev YuB, Skryabin AM, Pogodin RI. Inhalation of the aerosol of nuclear fuel particles from the Chernobyl nuclear power plant by adult persons from the Gomel region of Belarus. Environmental impact of radioactive releases. In: Proceedings of the IAEA International Symposium on Environmental Impact of Radioactive Releases; 8–12 May 1995; Vienna: IAEA; 1995. pp. 107-115

[41] Kutkov VA, Arefieva ZS, Muraviev YuB, Komaritskaya OI. Unique form of airborne radioactivity: Nuclear fuel "hot" particles of the Chernobyl accident. In: Proceedings of the IAEA international Symposium on Environmental Impact of Radioactive Releases; 8-12 May 1995; Vienna: IAEA; 1995. pp. 625-630

[42] ICRP Publication 119. Compendium of dose coefficients based on ICRP Publication 60. Annals of the ICRP. 2013;**42**(4):130

[43] ICRP-Database of Dose Coefficients: Workers and Members of the Public. Ver. 3.0. ICRP CD1. Available from: http://www.icrp.org [Accessed: 01-12-2018]

[44] ICRP Publication 88. Corrected version of May 2002. Doses to the embryo and fetus from intakes of radionuclides by the mother. Annals of the ICRP. 2001;**31**(1-2):511

[45] ICRP-Database of Dose Coefficients: Embryo and Fetus. Ver. 2.0. ICRP CD2. Available from: http://www.icrp.org [Accessed: 01-12-2018]

[46] ICRP Publication 95. Doses to infants from radionuclides ingested in mothers' milk. Annals of the ICRP. 2004;**34**(3-4):287

[47] ICRP-Database of Dose Coefficients: Radionuclides in Mother's Milk. Ver. 2.0. ICRP CD3. Available from: http://www.icrp.org [Accessed: 01-12-2018]

[48] Development of a Biokinetic Model for Radionuclide-Contaminated Wounds and Procedures for Their Assessment, Dosimetry and Treatment. NCRP Report 156. Bethesda, MD: NCRP; 2006. p. 428

[49] Toohey RE, Bertelli L, Sugarman SL, Wiley AL, Christensen DM. Dose Coefficients for Intakes of Radionuclides via Contaminated Wounds. Oak Ridge: ORISE; 2015. p. 628

Section 2

Radiation Therapy

Chapter 7

Spinal Stereotactic Body Radiotherapy (SBRT) Planning Techniques

Jina Kim, Yunji Seol, Hong Seok Jang and Young-Nam Kang

Abstract

Stereotactic body radiotherapy (SBRT) delivers a highly conformal and hypofractionated radiation dose to a small target with minimal radiation applied to the surrounding areas. The spine is an ideal site for SBRT owing to its relative immobility, the potential clinical benefits of high-dose delivery to this area, and the presence of adjacent critical structures such as the spinal cord, esophagus, and bowel. However, with the potential for radiation myelopathy if the dose is delivered inaccurately or if the spinal cord dose limit is set too high, proper treatment planning techniques for SBRT are important. Intensity modulation techniques are useful for spinal SBRT because of a rapid dose falloff and spinal cord avoidance. In this chapter, various planning techniques will be discussed and reviewed.

Keywords: SBRT, spine, IMRT, IMAT, tomotherapy, CyberKnife

1. Introduction

Stereotactic body radiotherapy (SBRT) was developed using the concepts of stereotactic radiosurgery (SRS). SRS was conceived by neurosurgeons and physicists in Sweden to allow the delivery of radiation to precise targets in the brain while minimizing injury to adjacent areas. The procedure delivers a high dose of radiation to the target accurately focused using multimodality imaging, such as computed tomography (CT), magnetic resonance imaging (MRI), and positron emission tomography/CT (PET/CT). The total dose is divided into several smaller doses of radiation, administered on separate days of treatment, typically in a single fraction or a few fractions. SRS treats tumors by destroying and distorting the DNA of these cells, in the same way as other forms of radiotherapy. As a result, these cells lose their ability to reproduce and die. Applied to the treatment of body tumors, the technique is called SBRT [1–4].

SBRT is also known as stereotactic ablative radiotherapy (SABR). SBRT ablates tumors by delivering precise and intensive radiation, guaranteeing minimal normal tissue complications. The characteristics of SBRT are summarized as follows: (1) a limited number of high dose-per-fraction treatments with a biologically equivalent dose (BED) of at least 75–100 as a minimum or even higher; (2) fields only slightly larger than gross tumor volume (GTV) with high accuracy even for moving targets, including the entire target with margins of 0.5–1.0 cm (i.e., exact delivery to tumor targets, sparing normal tissue); (3) dosimetry constructed to be very conformal,

IntechOpen

with sharp gradients from high- to low-dose areas; and (4) secure patient fixation during treatment and accurate duplication of patient position between simulation and treatment [2, 5–8].

Because of the high dose in a single fraction or fewer than five fractions, organs at risk (OARs) can be greatly affected by slight positional errors. Therefore, positional errors should be minimized. The margins of expansion can be reduced through the immobilization and control of respiratory motion of patients. Various commercial treatment delivery units in conjunction with the immobilization and respiratory motion control systems are available for the delivery of SBRT.

SBRT is currently both in use and being investigated for use in treating malignant or benign small- to medium-sized tumors in the body and at common disease sites, including the head and neck, lung, liver, abdomen, spine, and prostate. In particular, up to 70% of patients with malignancies are found to have skeletal involvement on postmortem examination, with the spine being the most common location [9]. For the treatment of spinal tumors, an extremely rapid dose falloff between the vertebral body and the spinal cord should be achieved [10, 11]. Implementation of correct beam-shaping and image-guided techniques has improved SBRT safety margins as well as accuracy and efficiency while accurately meeting 3D tumor contours. Spinal SBRT demands the highest accuracy in dose placement. In addition to patient fixation and multi-image guidance, a sophisticated treatment planning system that accurately models highly modulated small field beams is an indispensable factor in achieving high accuracy of radiation delivery.

To achieve this high accuracy, appropriate treatment planning technique should be used. Therefore, we will discuss various planning techniques for spinal SBRT in this chapter.

2. Spinal stereotactic body radiotherapy

2.1 Spine

The spine is a frequent site of metastases from primary cancer of the prostate, lung, breast, and kidney. After the lung and liver, the skeletal system is the most frequent site of metastases [12, 13], and 30% of all patients with cancer develop bone metastases [12, 14, 15]. In particular, bone metastasis occurs in 85% of patients with breast, prostate, and bronchial carcinoma [12, 16]. Approximately 50% of all bone metastases occur in the spinal cord. Of these, 60–80% are located in the thoracic spine, followed by 15–30% in the lumbar spine and less than 10% in the cervical spine [12, 13].

If left untreated, spinal metastases can cause axial pain, vertebral body fractures, radiculopathy, and the debilitating complications of metastatic epidural spinal cord compression (MESCC) [9]. The major complications of spinal metastases include neurologic dysfunction [12, 17, 18] and potential hypercalcemia, reduced activity, and bone fractures, resulting in a reduced quality of life [12, 16].

In general, primary spinal tumors are treated surgically, with the goal of maximal tumor removal. Numerous important blood vessels and adjacent organs surround the vertebrae. In particular, the spinal cord located in the vertebrae is a part of the central nervous system, which includes sensory and motor nerves. Complete resection of a tumor while preserving the nerve function of the spinal cord is difficult. In addition, vertebral instability due to tumor destruction or complete resection of the tumor must be considered, and fusion or fixation is often required for stability of the vertebrae. Depending on the malignancy of the tumor or the difficulty of complete resection, the patient may be treated with radiotherapy.

2.2 Radiotherapy for spinal tumors

2.2.1 Conventional method

Traditional radiotherapy methods of treating spinal tumors use large field radiation to treat the entire pathological vertebra and to treat one or two vertebral bodies, generally above and below the disease. This practice prevents missing the tumor owing to the limitations of diagnostic imaging and localization. In addition, the irradiation field of this technique is large but safe in the volume of the normal tissues irradiated because of the low biological effectiveness.

Large field radiation for spinal metastases has been the standard approach with outcomes of ~30% complete pain response and ~70% any response. The main limitation of the dose prescribed by traditional radiation techniques was the spinal cord. Overdosing radiation to the spinal cord has the devastating consequence of radiation-induced myelopathy that can leave the patient paralyzed. In addition to radiation myelopathy, possible toxicities include vertebral compression fractures and pain flares. Owing to the limitations of technology to prevent overdosing, clinical trials of high-dose effects on spinal metastasis have not been possible [19].

2.2.2 Stereotactic body radiotherapy

To overcome the limitations of conventional radiotherapy for the spine, hypofractionated treatment has been proposed, to deliver a high dose per fraction (typically 10–20 Gy/fraction), in contrast to the conventional fractionated treatment (2 Gy/fraction). The cumulative BED is significantly higher than that received in conventional treatment. Accurate delivery is of utmost importance owing to the high fractional dose and a small number of fractions. The delivery of an ablative dose to the target and rapid falloff doses away from the target enables minimization of the treatment toxicity to a tolerable level [20, 21]. In addition, there are other characteristics that distinguish SBRT from conventional radiotherapy, such as the number of beams used for treatment, the frequent use of non-coplanar beam arrangements, small or no beam margins on the penumbra, and the use of inhomogeneous dose distributions and dose-painting techniques 'including IMRT'. All of these technology improvements result in the highly conformal dose distribution that characterizes the SBRT technique [2].

Hypofractionated spinal SBRT has been shown to effectively and rapidly alleviate pain and improve neurological function in patients with or without epidural cord compression. SBRT allows minimal radiation exposure outside the target; the most significant problem associated with this procedure is related to spinal cord dose tolerance. Depending on the vertebral level of spinal metastasis, adjacent organs should be considered OARs. The tolerance of OARs to radiation from conventional fractionated radiotherapy is based on the entire organ or on a considerably large irradiated volume. SBRT delivers a highly conformal, hypofractionated radiation dose to a small target with minimal exposure of the surrounding areas to radiation [22].

A new radiotherapy technology that allows for intensity-modulated radiotherapy (IMRT) has emerged with spinal SBRT. IMRT is a technique designed to deliver a high biologically effective dose only to tumors within the vertebra for the purpose of tumor regression through permanent local control. The technique allows radiation beams to avoid the spinal cord, and even though a high dose is delivered to tumors, the dose received by the spinal cord is below the toxic threshold dose [23]. More details will be discussed in Section 3.

Table 1 lists maximum dose limits to a point or volume within several critical organs recommended for SBRT in one fraction (refer to TG-101 for

Serial tissue	Max critical volume	Max dose in critical volume (Gy)	End point (≥ Grade 3)
Spinal cord	<0.035 cc	14 Gy	Myelitis
	<0.35 cc	10 Gy	
	<1.2 cc (SBRT only)	7 Gy (SBRT only)	
Cauda equina	<0.035 cc	16 Gy	Neuritis
	<5 cc	14 Gy	
Sacral plexus	<0.035 cc	18 Gy	Neuropathy
	<5 cc	14.4 Gy	
Esophagus*	<0.035 cc	16 Gy	Stenosis/fistula
	<5 cc	11.9 Gy	
Ipsilateral brachial plexus	<0.035 cc	17.5 Gy	Neuropathy
	<3 cc	14 Gy	
Heart/pericardium	<0.035 cc	22 Gy	Pericarditis
	<15 cc	16 Gy	
Great vessels*	<0.035 cc	37 Gy	Aneurysm
	<10 cc	31 Gy	
Trachea* and larynx	<0.035 cc	20.2 Gy	Stenosis/fistula
	<4 cc	10.5 Gy	
Skin	<0.035 cc	26 Gy	Ulceration
	<10 cc	23 Gy	
Stomach	<0.035 cc	16 Gy	Ulceration/fistula
	<10 cc	11.2 Gy	
Duodenum*	<0.035 cc	16 Gy	Ulceration
	<5 cc	11.2 Gy	
Jejunum/ileum*	<0.035 cc	15.4 Gy	Enteritis/obstruction
	<5 cc	11.9 Gy	
Colon*	<0.035 cc	18.4 Gy	Colitis/fistula
	<20 cc	14.3 Gy	
Rectum*	<0.035 cc	18.4 Gy	Proctitis/fistula
	<20 cc	14.3 Gy	
Renal hilum/vascular trunk	<2/3 volume	10.6 Gy	Malignant hypertension
Parallel tissue	**Critical volume (cc)**	**Max dose in critical volume (Gy)**	**End point (≥Grade 3)**
Lung (right and left)	1000 cc	7.4 Gy	Pneumonitis
Renal cortex (right and left)	200 cc	8.4 Gy	Basic renal function

**Avoid circumferential irradiation.*

Table 1.
One fraction dose constraints of several critical organs from RTOG 0613 [24].

multiple-fraction dose constraints [2]). The recommended dose constraints are shown in max critical volume and the maximum dose to the given volume for each organ. These limitations have been determined based on the widely accepted radiosurgery norms currently in practice. Regardless of these limitations, the participating centers are encouraged to adhere to the prudent treatment planning principle to avoid unnecessary radiation exposure to critical normal structures [24].

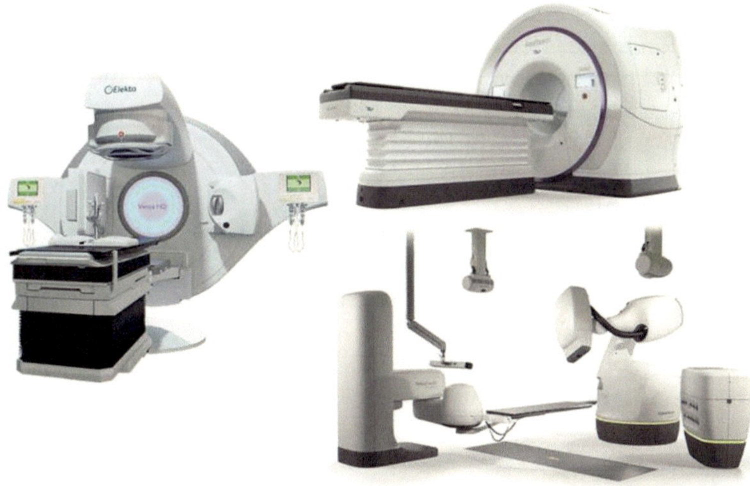

Figure 1.
Commercial treatment delivery units. From left to right are versa HD (Elekta AB), Radixact (Accuray Inc.), and CyberKnife M6 (Accuray Inc.).

2.2.3 SBRT delivery systems

Various commercial treatment delivery units can be used to deliver SBRT [1, 5, 7], as shown in **Figure 1**. They all have the capability of image-guided radiotherapy, enabling tumor or target localization prior to treatment delivery and allowing treatment setup uncertainty to be significantly reduced. All delivery units, with the exception of proton therapy, used as photon-based SBRT, are linear accelerators (LINACs). There are several types of image-guidance equipment: 2D imaging types, including room-mounted or gantry-mounted orthogonal kilovoltage (kV) radiographs and fluoroscopy, and 3D imaging types including kV or megavoltage (MV) cone-beam CT (CBCT) and CT-on-rails in room.

In addition to general LINACs, there are many types of treatment systems. The CyberKnife (CK, Accuray Inc., Sunnyvale, CA, USA) unit has a six-axis robotic manipulator that enables delivery of the beam to the target from many different directions in order to minimize radiation exposure to nearby organs. A pair of orthogonally positioned imaging systems enables monitoring of the target motion, with automatic correction. CK is a commonly used modality for SBRT owing to its highly conformal dose distributions, steep gradient, and near real-time image-guidance system. The helical tomotherapy (HT, Accuray) unit is a special device performing continuous 360° rotations using a binary multi-leaf collimator (MLC), with the treatment couch moving continuously during the treatment [1].

Each treatment delivery system has strengths and weaknesses. An appropriate treatment delivery system and corresponding optimal planning technique should be used for successful and safe treatment.

3. SBRT planning techniques

SBRT is a high-precision radiotherapy technique that utilizes the high doses of radiation in a single fraction or a few fractions, as mentioned in the above sections. In principle, three-dimensional conformal radiotherapy (3D-CRT) planning can be applied to SBRT. When the beams at multiple angles are concentrated at the center of small lesions, a high-dose heterogeneity that contributes to a steep dose gradient at the target edge appears and may be desirable in terms of normal tissue sparing and dose escalation to the GTV [1].

To treat a spinal tumor, conventionally fractionated 3D-CRT modifies the beam shape to match the projection of target volume at each gantry angle using an MLC. The accuracy of the shape of the beam projected onto the target depends on the width of leaves. MLC leaf widths of 2.5–10 mm have been reported for use in SBRT planning [25, 26].

However, delivery to the target is limited by tolerance of normal tissues, particularly the spinal cord, so it is necessary to irradiate the target with lower dose. In suboptimal cases, several side effects can occur, such as paraplegia, pain, increased steroid use, and reduced survival rate.

3.1 Intensity-modulated radiation treatment

The development of IMRT was a major improvement over 3D-CRT for SBRT [27]. IMRT allows for the radiation dose to conform more precisely to the shape of the tumor by modulating the intensity of the radiation beam and allows higher radiation doses to be focused to regions within the tumor, sparing the surrounding normal critical structures. In particular, when treating spinal tumors, intensity modulation allows production of a concave-shaped dose distribution with the exception of the spinal cord.

The IMRT technique uses computerized inverse planning. Conformal radiotherapy is forward planning and depends on the skills of the treatment planner to determine the number, shape, and orientation of the beams. Inverse planning, in contrast, specifies the plan outcome in terms of the tumor dose and normal structure dose limits. The computer system then adjusts the beam intensities to identify a configuration best matched to the desired plan [28].

During the procedure, each beam is divided into several beam elements (beamlets) of a few millimeters, and the relative weight is optimized so that the desired dose distribution appears. The optimization process involves inverse planning in which beamlet weights or intensities are adjusted to satisfy predefined dose criteria for the composite plan. When optimization is complete, an optimized fluence map generates a sequence of MLC leaves for each beam. The field at one gantry angle is subdivided into a set of subfields irradiated at a uniform beam intensity level. The subfields are shaped by the MLC, and the intensity-modulated field is obtained by summing several subfields.

The two most common methods of IMRT delivery are segmental (step-and-shoot) and dynamic (sliding window). The difference between the two is the motion of MLC at a given gantry angle. In segmental MLC delivery, the beam is turned off while the leaves move until the next subfield is prepared. The advantage of the segmental MLC method is that it is easy to plan and no additional dose can occur while the MLC is moving to create the next subfield. On the other hand, the dose delivery is slow owing to the delay in turning the beam on and off, resulting in an increase in treatment time. In the dynamic MLC delivery, the MLC leaves are moving during irradiation. Each pair of leaves sweeps across target volumes under computer control. Dynamic MLC delivery offers better dose homogeneity for target volume and shorter treatment time in comparison to the segmental MLC; however, the larger total irradiated dose is a disadvantage.

Compared to 3D-CRT, the dose distribution can be made even more sophisticated because target coverage and avoidance of critical structures located adjacent to the target volume are better. The more sophisticated implementation of SBRT has become possible with the IMRT technique. The technique mentioned in this section (Section 3.1) was the IMRT technique with a fixed gantry, and IMRT with a rotating gantry will be discussed in the following section (Section 3.2).

3.2 Intensity-modulated arc therapy

Intensity-modulated arc therapy (IMAT) is a combined technique of IMRT and rotational treatment. When performed for a C-shaped target with a sensitive structure in the concavity of the "C," like a spinal tumor, the rotational treatment has a dosimetric advantage. The result of simulation that supports this is that when all the planning parameters except the beam angle number are constant, the dose becomes more homogeneous in the tumor and decreased in the critical structures as the number of angles increased [29].

IMAT uses rotational cone beams of varying aperture shapes and varying dose weightings to achieve intensity modulation. However, the speed of rotation cannot have frequent and drastic variations owing to the weight of the LINAC gantry; therefore, the variations in dose weighting are primarily achieved through varying the machine dose weight. MLC moves dynamically to shape each subfield while the gantry is rotating and the beam is on continuously [30]. Arcs are approximated as multiple-shaped fields in a regular angular interval. One subfield is delivered at each arc. The next new arc is started to deliver the next subfield and so on until all the planned arcs and their subfields have been delivered. That is, overlapping arcs create intensity modulation.

To create more effective treatment plans, various techniques have been purposed within IMAT. Volumetric-modulated arc therapy (VMAT) and modulated arc therapy (mARC) are examples of such techniques. VMAT is a single or multi-arc form of IMRT technique that changes the dose rate and gantry speed while the gantry is rotating. Currently there are several VMAT systems available under various names (RapidArc, Varian Medical Systems, Palo Alto, CA, USA; SmartArc, Philips Radiation Oncology Systems, Fitchburg, WI, USA; and Elekta VMAT, Elekta, Stockholm, Sweden) [31]. The mARC technique as an alternative to VMAT is a rotational IMRT irradiation with burst mode delivery. Both the dose rate and gantry speed are modulated to allow for delivery of the correct dose per IMRT segment, and an MLC velocity servo is required to continuously adjust the leaf velocity to facilitate accurate, and timely, leaf positioning [32].

The technique is similar to HT, which is an IMRT technique that rotates in a helical form and will be discussed in Section 3.3. As compared with HT, IMAT has certain advantages: (1) IMAT eliminates the need for transferring the patient during treatment and avoids abutment issues as seen with serial HT, (2) IMAT retains the ability to use non-coplanar beams and arcs, and (3) IMAT uses a conventional LINAC; thus, complex rotational IMRT treatments and simple palliative treatments can be delivered with the same treatment unit [30].

The main advantages of rotational therapy compared to fixed-gantry IMRT are improved conformity of the dose distribution in the high-dose regions, as well as possible reduction of the treatment time. The short treatment time can lead to improved patient comfort and reduce the risk of movement. Moreover, shorter treatment times can be biologically beneficial. Radiation survival is not only a function of the total dose delivered but also depends on the duration of radiation delivery [33, 34]. IMAT offers the efficient use of monitor units (MUs). The number of MUs per treatment is correlated with the amount of scatter dose and leakage radiation, which could be important in view of the induction of secondary malignancies [35]. The decrease in MUs achieved with IMAT partly addresses this issue, which is one of the major concerns with IMRT [36].

However, the complex nature of IMAT planning has been one of the primary barriers to routine clinical implementation. From one angle to the next in each VMAT arc, leaf motion between adjacent angles is limited by leaf travel speed and gantry rotation speed. Therefore, the technique has disadvantages such as difficulty and complexity of planning.

3.3 Helical tomotherapy

HT is a radiotherapy modality that combines helical CT scanning with an MV linear accelerator. A 6 MV LINAC rotates on a ring gantry at a source-axis distance (SAD) of 85 cm, and the beam passes through a primary collimator into a fan-beam shape. During treatment, the ring gantry continuously rotates, while the couch is continuously translated through the rotating beam plane. The dose is thus delivered in a helical fashion. The ring gantry also contains a detector system that is mounted opposite the accelerator and is used to collect data for megavoltage CT (MVCT) acquisition. A beam stopper is used to reduce room-shielding requirements [37].

The MVCT in HT is used as a tool to enhance image-guided daily treatment setup and positioning of the patient. Because SBRT usually requires a longer treatment period owing to the use of high-dose hypofraction, the patient must be fixed in place to limit the patient's movement during treatment. However, patients with vertebral metastases, in particular, often move involuntarily during treatment owing to back pain that cannot be controlled. Therefore, it is important to ensure the accuracy of high-dose delivery and to avoid side effects of OARs on intrafractional movement. A daily MVCT image scan is generated prior to treatment to ensure accurate delivery of each treatment according to the patient's anatomy on a particular day. This MVCT is integrated with the kilovoltage CT (kVCT) imaging plan to provide a reference for patient setup and positioning [38].

The fan-beam has an extension of 40 cm in the lateral direction and smaller or equal to 5 cm (typically 1.0, 2.5, and 5.0 cm) in the longitudinal direction at the isocenter. With the use of a compressed air-driven multi-leaf (64 leaves) binary collimator (MLC), radiation beams are shaped, and their intensities are modulated. The leaves are mounted on two opposite blocks, and each individual leaf is driven from open to closed state. The intensity modulation is achieved by controlling the length of time each leaf is open. Each leaf has a width of 6.25 mm (40 cm divided by 64 leaves) and rapid transitioning (about 20 ms); thus it can produce a sufficiently accurate shape even within a short rotation period. Therefore, HT offers a very useful treatment modality of spinal SBRT by implementing image-guided radiation therapy (IGRT) and IMRT techniques.

For the treatment planning of each rotation, a rotation is divided into 51 projections (360°/7° = 51). For each projection, each MLC leaf has a unique opening time as shown in **Figure 2** [39]. Unlike the usual LINAC radiotherapy, there are additional parameters: slice width, pitch factor, and modulation factor. These parameters influence both treatment time and quality of the treatment plan.

Slice width (or field width) is the longitudinal extent (i.e., in the y-direction) of the treatment field. For planning purposes, a nominal 1.0, 2.5, or 5.0 cm is selected. Pitch is defined as distance traveled by the couch per gantry rotation, divided by the slice width. With a lower pitch value, there is greater overlap between spirals. This factor influences the treatment time. Modulation factor is defined as the maximum leaf opening time divided by the average opening time of all leaves. This value can range from 1.0 to 10 (typically using from 1.5 to 3.5). For a complex treatment requiring a lot of MLC motion, a high modulation factor is selected.

One of the most important differences between the HT system and other radiotherapy systems is that the HT system does not have a flattening filter. The main advantages of an absent flattening filter are an increased dose rate, reduced scatter, reduced leakage, and reduced out-of-field doses [40, 41]. The main reason for allowing the nonuniform profile is that HT is a dedicated IMRT system, without the need for a flat dose profile. If it is still desired, the MLC can be used to modulate the treatment field to produce a flat dose distribution [42].

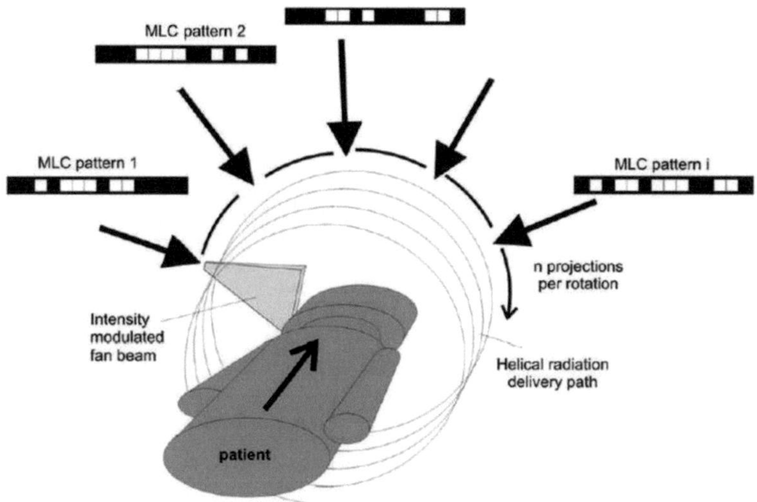

Figure 2.
Illustration of the helical tomotherapy delivery. Copyright © Journal of Medical Physics.

In treating spinal tumors, the major requirement is minimization of the dose to the spinal cord. The dose gradient should be increased to improve the conformity while allowing increased heterogeneity in the tumor volume coverage. In addition, the slice width and pitch parameters are considered to increase cord avoidance and target coverage.

3.4 CyberKnife

CK is one of the representative delivery units of SBRT. As mentioned briefly in the above section, CK has uniquely different features compared with the common medical LINACs. The compact LINAC mounted on a computer-controlled six-axis robotic manipulator delivers radiation beams anywhere in the body with submillimeter accuracy. The integrated orthogonally positioned kV X-ray imaging system is utilized to monitor the patient position throughout the course of radiotherapy. Patients are positioned automatically or manually by a therapist by matching fiducial markers or bony anatomy from X-ray images to digital reconstructed radiographs generated by CT simulation [43].

The robotic manipulator with six degrees of freedom can deliver the beam anywhere in space. Accordingly, the beam position and orientation can be adjusted by the robot to accommodate changes in target position and orientation during treatment without the need to move the patient.

The beam field size is controlled through various collimation types: 12 fixed cone collimators or an Iris variable collimator (Accuray) consisting of 12 tungsten leaves that produce beam diameters ranging from 5 to 60 mm (defined at 800 mm distance from the X-ray source) [44]. Furthermore, to compensate for the limit caused by the fixed field size, an MLC has recently been introduced for the CK [45]. The new MLC system consists of 41 leaf pairs, each with a width of 2.5 mm. The maximum field size is 12 × 10.25 cm. This new system allows the fields to be shaped matching the tumor shape and allows reduction of treatment time. In particular, using the MLC offers a dosimetric advantage for targets near OARs, as shown in **Figure 3** [46].

The unit delivers multiple isocentric or non-isocentric photon beams to a desired target from many different angles through a robotic arm, as well as optic image guidance for motion management. The isocentric treatment planning is similar to

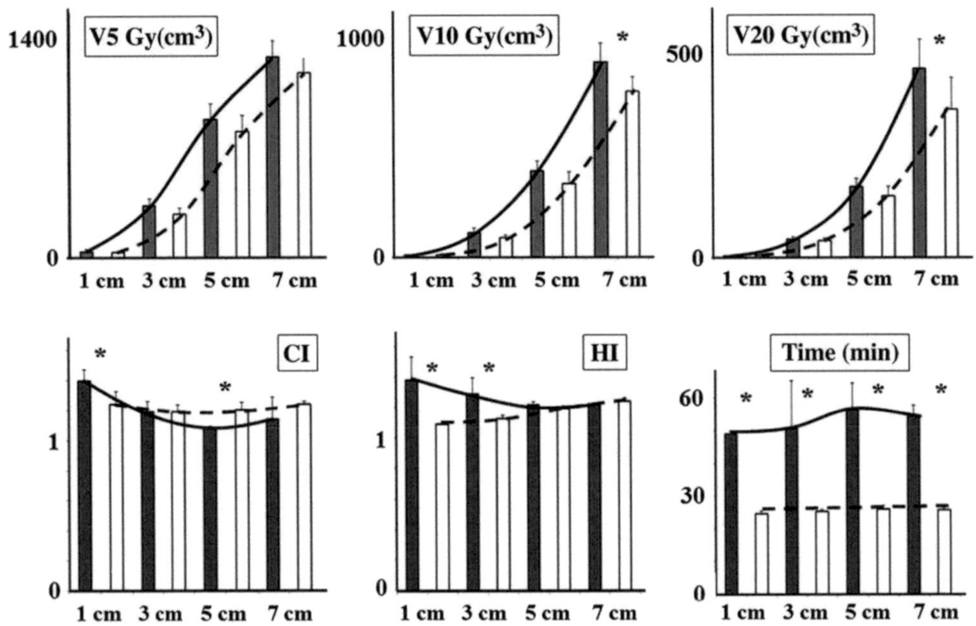

Figure 3.
Dose-volume parameters in circular collimator and multi-leaf collimator (MLC) plans for 1–7 cm brain target volumes. (White bars indicate multi-leaf collimator (MLC) and gray indicate circular collimator.) Copyright © 2017, Oxford University Press.

that of the Gammaknife (Elekta) and conventional LINACs, which have a fixed mechanical center of the gantry and collimator. The location of the isocenter is not limited, providing a great advantage over many other delivery units. However, this advantage can be overcome by using inverse planning; the final target dose distributions can be manipulated to a certain level by modifying the order of the targets as well as the contours and dose limits assigned to the target and critical organs.

In non-isocentric treatment planning, radiation beams are delivered to a specific portion of the tumor without couch repositioning. This technique makes the high-dose isodose lines match the target shape and avoid nearby critical organs. Therefore, non-isocentric planning is very useful for treatment of irregularly shaped targets. CK, which is available with both plans, is advantageous for combining the rapid dose falloff of isocentric plans with the dose conformity of non-isocentric plans [1].

3.5 Planning considerations

In spinal SBRT, the target volume includes the involved vertebral body and both left and right pedicles and the grossly visible tumor, if a paraspinal or epidural lesion is present. The target volume is generally delineated with no margin. However, depending on the treatment system, a beam aperture margin of 2–3 mm beyond the target volume is allowed to ensure adequate dose coverage of the target. This margin can be reduced to 0–1 mm in the area of the spinal cord to meet spinal cord dose constraints. The target volume may be selected at the discretion of the treating radiation oncologist based on the extent of tumor involvement. In any circumstance in which there is an epidural or paraspinal soft tissue tumor component, the visible epidural or paraspinal tumors are included in the target volume [24].

Normal tissue contouring is required starting at 10 cm above the target volume to 10 cm below the target. The treatment plan should be established according to the recommended maximum dose limit for several critical organs, as shown in **Table 1**.

Among the dose-limiting critical OARs, the spinal cord is a key concern. Because of the nature of radiosurgery with a rapid dose falloff, there is a radiation dose gradient within the diameter of the spinal cord. Therefore, a partial spinal cord volume defined as from 5 to 6 mm above to 5–6 mm below the target volume is used. The partial or absolute volume spinal cord constraints are applied to each treated spine level when the patient has multiple spine levels treated. Any spinal cord dose exceeding this constraint is not acceptable and is a major deviation [24].

Successful treatment planning requires 90% coverage of the target volume by the prescribed dose. Typically, the 80–90% isodose line is used as the prescription line, although the prescription isodose line may be different depending on the delivery system. Coverage of <90% of the target volume is an acceptable variation, and any coverage of <80% of the target volume is an unacceptable deviation. The treatment plan is acceptable as long as ≥90% of the target volume receives the prescribed dose. It should be noted, however, that owing to the irregular shape of the target volume and the location of the spinal cord, hot spots may be created in the immediate vicinity outside of the target volume [24].

Because of the characteristics of the spinal SBRT, in the case of a beam with a small size, the higher the beam energy, the larger the beam penumbra as a result of lateral electron transport in the medium. The commonly available 5 mm MLC leaf width has been found to be adequate for most applications, with negligible improvements using the 3 mm leaf width MLC for all but the smallest lesions (<3 cm in diameter). A 6 MV photon beam, available on most modern treatment machines, provides a reasonable compromise between the beam penetration and penumbra characteristics. Additionally, beam arrays should be placed mostly in the posterior direction to avoid entrance of the radiation beam through the lungs. In the case of arc rotation techniques, every effort should be used to limit the passage of radiation through the lungs [2].

4. Comparison of plan result

In Section 3, several spinal SBRT planning techniques were discussed. Because the planning technique should be selected depending on the patient's condition or situation, numerous studies have been performed to compare various planning techniques for treating spinal tumors. To evaluate the results of each plan for spinal SBRT, the following quantitative parameters were used [22, 47–50].

- Conformity index (CI): a measure of the dose coverage to the planned target volume (PTV).
- Dice similarity coefficient (DSC): a spatial overlap index and a reproducibility validation metric [51].
- Homogeneity index (HI): a measure of uniformity of the dose within the target volume.
- PTV coverage: 100% of the PTV receiving the prescribed dose [52].
- Spinal cord dose: maximum dose to the spinal cord.
- High-dose spillage: The cumulative volume of all tissue outside the PTV receiving a dose >105% of prescription dose should be no more than 15% of the PTV volume [53].

- Intermediate-dose spillage (R50% and D2cm): the falloff gradient located outside of the PTV.

  R50%: volume that received 50% of the prescribed dose/PTV volume

  D2cm: maximum dose in terms of the percentage of the prescribed dose at 2 cm beyond the PTV in any direction

- Equivalent uniform dose (EUD): the absorbed dose that, if homogeneously delivered to a tumor, causes the same expected number of clonogens to survive as does the actual nonhomogeneous absorbed dose distribution.

- Biological effective dose (BED): the dose producing equivalent biological effect regardless of dose uniformity or fractionations.

- Gamma index: the standard method for planar dose verification in IMRT QA; calculates the quantity γ for each point of interest using preselected dose difference (DD) and distance to agreement (DTA) criteria and then uses the γ value to determine the outcome (pass-fail) of the IMRT QA [53].

In addition, plans were evaluated by the treatment delivery time (beam irradiated time) or the target point dose for the phantom measured in the ion chamber.

Zach et al. compared VMAT to static beam IMRT for spinal SBRT. The plans were compared for conformity, homogeneity, treatment delivery time, spinal cord dose, and D_{max} of the spinal cord and V 10 Gy, which is the volume of the spinal cord exposed to at least 10 Gy. The authors also compared the monitor units required in each plan to compute the net irradiated time.

All evaluated parameters were shown to favor the VMAT plans over the IMRT plans. D_{min} for PTV in the IMRT was significantly lower than that in the VMAT plan. The DSC and treatment time were found to be significantly better for the VMAT plans than for the IMRT plans. A reduction of almost 50% in the net treatment time was calculated. The authors reported that VMAT provides better conformity, homogeneity, and spinal cord dose. They also suggested that the shorter treatment time is a major advantage and not only provides convenience for patients experiencing pain but also contributes to the precision of this high-dose radiotherapy [47].

In another study, Choi et al. compared the treatment planning performance of RapidArc (i.e., VMAT) and CK for spinal SBRT. The optimized dose priorities for both plans were similar for all patients. The highest priority was to provide sufficient dose coverage to the PTV while limiting the maximum dose to the spinal cord. Plan quality was evaluated with respect to PTV coverage, CI, high-dose spillage, intermediate-dose spillage, and maximum dose to the spinal cord, which are criteria recommended by the RTOG 0631 spine and 0915 lung SBRT protocols.

The mean CI ± standard deviation (SD) values of the PTV were 1.11 ± 0.03 and 1.17 ± 0.10 for RapidArc and CK, respectively. On average, the maximum dose delivered to the spinal cord in CK plans was approximately 11.6% higher than that in RapidArc plans. High-dose spillages were 0.86 and 2.26% for RapidArc and CK, respectively. Intermediate-dose spillage characterized by D2cm was lower for RapidArc than for CK; however, R50% was not statistically different between the plans. Although both systems can create highly conformal volumetric dose distributions, the study of Choi et al. shows that RapidArc was associated with lower high- and intermediate-dose spillages than was CK. The authors also suggested that RapidArc plans for spinal SBRT may be superior to CK plans [48].

Sahgal et al. compared the treatment planning quality of the CK and Novalis (BrainLAB AG, Heimstetten, Germany) systems for vertebral body SBRT. Physical parameters and biological modeling parameters such as PTV dose coverage, dose conformity, EUD, integral BED, and a generalized BED were used to compare the treatment plans.

In the study, both the CK and Novalis treatment plans fulfilled the specified requirements with comparable PTV dose coverage and dose conformity. For the target volume, CK plans produced significantly higher values of all calculated parameters to the PTV. For OARs, CK plans produced a somewhat lower dose to small volumes (0.1–1 cm^3) of the spinal cord and esophagus but exposed larger volumes of these structures to a low dose as compared to the Novalis plans.

The authors reported that restricting the dose to a small volume of the spinal cord and esophagus resulted in a modest decrease in the dose to 1 cm^3 volume of these structures for CK planning but at the expense of a larger volume of these structures exposed to low-dose levels [49].

In another study, Kim et al. compared the planning characteristics for hypo-fractionated spinal SBRT administered using three treatment techniques (IMRT, mARC, and HT). The factors evaluated for spinal SBRT planning were dose coverage, cord avoidance, target conformity, homogeneity, and dose spillage.

Target dose coverage was 82.74 ± 3.35, 80.92 ± 0.81, and 85.01 ± 7.27% for IMRT, mARC, and HT, respectively. The authors reported that HT was therefore a powerful technique with respect to target coverage. The spinal cord dose for HT (mean, 1763.96 cGy; SD, 164.48) was significantly different from those for mARC (mean, 1991.75 cGy; SD, 248.00) and IMRT (mean, 2053.24 cGy; SD, 164.48). In addition, the partial spinal cord volume at 2000 cGy for HT (mean, 0.12 cc; SD, 0.01) was significantly different from those for IMRT and mARC (0.50 ± 0.10 cc and 0.56 ± 0.25 cc, respectively). The CIs were 1.30 ± 0.12, 1.08 ± 0.05, and 1.36 ± 0.23 for IMRT, mARC, and HT planning, respectively. mARC showed the highest conformity. Regarding HI, HT (mean, 1763.96 cGy; SD, 164.48) differed statistically from both mARC (mean, 1991.75 cGy; SD, 248.00) and IMRT (mean, 2053.24 cGy; SD, 164.48) with respect to the spinal cord dose.

HT used a narrow field fan-beam and exhibited remarkable improvement of target coverage and cord dose, offering an important benefit to spinal SBRT. mARC had the highest target conformity and showed more favorable high- and intermediate-dose spillage than did HT and IMRT. These three planning techniques have different advantages. The authors suggested utilizing different planning techniques according to the cases. In the case of spinal SBRT, HT should be used for cord avoidance. In some cases, such as for a short treatment duration when the patient is considered to be in poor general condition, mARC can be used [22].

Gallo et al. performed end-to-end (E2E) testing for a set of representative spinal targets planned and delivered using four different treatment planning systems and delivery systems, specifically HT, Vero, TrueBeam with flattening filter free (FFF) and flattened, and CK, to evaluate the various capabilities of each. An anthropomorphic E2E SBRT phantom was simulated and treated on each system to evaluate agreement between measured and planned doses. The phantom accepted 0.007 cm^3 ion chambers in the thoracic region and radiochromic film in the lumbar region.

Ion chamber measurements in the thoracic targets resulted in an overall average difference of 1.5% with planned doses. Specifically, measurements agreed with the treatment planning system to within 2.2, 3.2, 1.4, 3.1, and 3.0% for all three measureable cases on HT, Vero, TrueBeam (FFF), TrueBeam (flattened), and CK, respectively. Film measurements for the lumbar targets resulted in average global gamma index passing rates of 100 at 3%/3 mm, 96.9 at 2%/2 mm, and 61.8 at 1%/1 mm, with a 10% minimum threshold for all plans on all platforms. Local gamma analysis was also performed with similar results. While gamma passing rates were consistently accurate across all platforms through 2%/2 mm, treatment beam-on delivery times varied greatly among the platforms, with TrueBeam (FFF) the shortest, averaging 4.4 min, TrueBeam using flattened beam at 9.5 min, HT at 30.5 min, Vero at 19 min, and CK at 46.0 min.

In the study, despite the complexity of the representative targets and their proximity to the spinal cord, all treatment platforms were able to create plans that meet all RTOG 0631 dose constraints and produced exceptional agreement between calculated and measured doses. However, there were differences in the plan characteristics and significant differences in the beam-on delivery time between platforms. Thus, the authors stated that clinical judgment is required in each particular case to determine the most appropriate treatment planning/delivery platform [50].

5. Conclusion

This chapter has described various planning techniques for spinal SBRT and summarized the studies comparing these techniques. The spine is a frequent site of tumor metastasis, but there are many important vessels and adjacent organs in the vicinity of the vertebrae. In particular, the spinal cord within the spine is part of the central nervous system. Radiotherapy is performed depending on the malignancy of the tumor or the difficulty of complete resection, considering potential spinal instability caused by the tumor destruction or complete resection. However, the major limitation of traditional radiotherapy is the tolerance dose of the spinal cord. If the spinal cord is irradiated with an overdose, toxicities such as radiation-induced myelopathy, vertebral compression fracture, or pain flare may occur. To overcome the limitation of conventional radiotherapy, SBRT has been proposed. The technique of SBRT delivers a higher BED, within the range of what is considered locally curative. A conformal high-dose beam in a few fractions should be used, and an intensity modulation technique is required for the sparing of normal organs surrounding the spinal lesion. Various planning technologies based on intensity modulation technology are available, including IMRT with fixed gantry, IMAT, HT, and CK. Different planning techniques have their distinct features and advantages. Therefore, it is important to use appropriate treatment planning depending on the patient's condition and situation.

Acknowledgements

This research was supported by Advanced Institute for Radiation Fusion Medical Technology (AIRFMT) at the Catholic University of Korea.

Conflict of interest

The authors report no conflicts of interest.

Author details

Jina Kim[1], Yunji Seol[1], Hong Seok Jang[2] and Young-Nam Kang[2*]

1 The Catholic University of Korea, Seoul, Republic of Korea

2 Department of Radiation Oncology, Seoul St. Mary's Hospital, College of Medicine, The Catholic University of Korea, Seoul, Republic of Korea

*Address all correspondence to: ynkang33@gmail.com

References

[1] Lo SS, Teh BS, Lu JJ, Schefter TE. Stereotactic Body Radiation Therapy. Berlin, Heidelberg, Germany: Springer; 2012

[2] Benedict SH, Yenice KM, Followill D, Galvin JM, Hinson W, Kavanagh B, et al. Stereotactic body radiation therapy: The report of AAPM task group 101. Medical Physics. 2010;**37**(8):4078-4101

[3] Song CW, Cho LC, Yuan J, Dusenbery KE, Griffin RJ, Levitt SH. Radiobiology of stereotactic body radiation therapy/stereotactic radiosurgery and the linear-quadratic model. International Journal of Radiation Oncology, Biology, Physics. 2013;**87**(1):18-19

[4] Bijlani A, Aguzzi G, Schaal D, Romanelli P. Stereotactic radiosurgery and stereotactic body radiation therapy cost-effectiveness results. Frontiers in Oncology. 2013;**3**:77

[5] Timmerman RD, Kavanagh BD, Cho LC, Papiez L, Xing L. Stereotactic body radiation therapy in multiple organ sites. Journal of Clinical Oncology. 2007;**25**(8):947-952

[6] Meyer J. IMRT, IGRT, SBRT: Advances in the Treatment Planning and Delivery of Radiotherapy. In: Meyer JL, editor. Karger Medical and Scientific Publishers. Switzerland: KARGER Publishers; 2011

[7] Kay C-S, Kang Y-N. Curative radiotherapy in metastatic disease: How to develop the role of radiotherapy from local to metastases. In: Frontiers in Radiation Oncology. United Kingdom: IntechOpen; 2013

[8] Potters L, Steinberg M, Rose C, Timmerman R, Ryu S, Hevezi JM, et al. American society for therapeutic radiology and oncology* and american college of radiology practice guideline for the performance of stereotactic body radiation therapy. International Journal of Radiation Oncology, Biology, Physics. 2004;**60**(4):1026-1032

[9] Tseng C-L, Eppinga W, Charest-Morin R, Soliman H, Myrehaug S, Maralani PJ, et al. Spine stereotactic body radiotherapy: Indications, outcomes, and points of caution. Global Spine Journal. 2017;**7**(2):179-197

[10] Yamada Y, Bilsky MH, Lovelock DM, Venkatraman ES, Toner S, Johnson J, et al. High-dose, single-fraction image-guided intensity-modulated radiotherapy for metastatic spinal lesions. International Journal of Radiation Oncology, Biology, Physics. 2008;**71**(2):484-490

[11] Yin FF, Ryu S, Ajlouni M, Zhu J, Yan H, Guan H, et al. A technique of intensity-modulated radiosurgery (IMRS) for spinal tumors. Medical Physics. 2002;**29**(12):2815-2822

[12] Schneider F, Greineck F, Clausen S, Mai S, Obertacke U, Reis T, et al. Development of a novel method for intraoperative radiotherapy during kyphoplasty for spinal metastases (Kypho-IORT). International Journal of Radiation Oncology, Biology, Physics. 2011;**81**(4):1114-1119

[13] Bartels RH, van der Linden YM, van der Graaf WT. Spinal extradural metastasis: Review of current treatment options. CA: A Cancer Journal for Clinicians. 2008;**58**(4):245-259

[14] Yuen KK, Shelley M, Sze WM, Wilt TJ, Mason M. Bisphosphonates for advanced prostate cancer. Cochrane Database of Systematic Reviews. 2006;(4):CD006250. https://www.cochranelibrary.com/cdsr/doi/10.1002/14651858.CD006250/full

[15] Pavlakis N, Schmidt RL, Stockler MR. Bisphosphonates for breast cancer.

Cochrane Database of Systematic Reviews. 2005;(3):CD003474. https://www.cochranelibrary.com/cdsr/doi/10.1002/14651858.CD003474.pub2/full

[16] Mercadante S. Malignant bone pain: Pathophysiology and treatment. Pain. 1997;**69**(1-2):1-18

[17] Sheehan JP, Jagannathan J. Review of spinal radiosurgery: A minimally invasive approach for the treatment of spinal and paraspinal metastases. Neurosurgical Focus. 2008;**25**(2):E18

[18] Jagas M, Patrzyk R, Zwoliński J, Kołodziejczyk A, Sakowski J, Pudełko M, et al. Vertebroplasty with methacrylate bone cement and radiotherapy in the treatment of spinal metastases with epidural spinal cord compression. Preliminary report. Ortopedia, Traumatologia, Rehabilitacja. 2005;7(5):491-498

[19] McDonald R, Chow E, Rowbottom L, DeAngelis C, Soliman H. Incidence of pain flare in radiation treatment of bone metastases: A literature review. Journal of Bone Oncology. 2014;**3**(3-4):84-89

[20] Zhang J, Yang F, Li B, Li H, Liu J, Huang W, et al. Which is the optimal biologically effective dose of stereotactic body radiotherapy for stage I non-small-cell lung cancer? a meta-analysis. International Journal of Radiation Oncology, Biology, Physics. 2011;**81**(4):e305-e316

[21] Onishi H, Araki T, Shirato H, Nagata Y, Hiraoka M, Gomi K, et al. Stereotactic hypofractionated high-dose irradiation for stage I nonsmall cell lung carcinoma: Clinical outcomes in 245 subjects in a Japanese multiinstitutional study. Cancer. 2004;**101**(7):1623-1631

[22] Kim J, Jang HS, Kim YS, Choi BO, Kang Y-N. Comparison of spinal stereotactic body radiotherapy (SBRT) planning techniques: Intensity-modulated radiation therapy, modulated arc therapy, and helical tomotherapy. Medical Dosimetry. 2017;**42**(3):210-215

[23] Sahgal A. Spine Stereotactic Body Radiotherapy. Hot Spot [Internet]. 2008; 10(4). Available from: https://sunnybrook.ca/uploads/HS081101.pdf

[24] Ryu S, Dicker A, Gerszten PC, Movsas B, Yin F-F, Wang M, et al. RTOG 0631 Phase II/III Study of Image Guided Stereotactic Radiosurgery/SBRT for Localized Spine Metastases. 2010. Available from: https://www.rtog.org/ClinicalTrials/ProtocolTable/StudyDetails.aspx?study=0631

[25] Nelson JW, Yoo DS, Sampson JH, Isaacs RE, Larrier NA, Marks LB, et al. Stereotactic body radiotherapy for lesions of the spine and paraspinal regions. International Journal of Radiation Oncology, Biology, Physics. 2009;**73**(5):1369-1375

[26] Tanyi JA, Summers PA, McCracken CL, Chen Y, Ku L-C, Fuss M. Implications of a high-definition multileaf collimator (HD-MLC) on treatment planning techniques for stereotactic body radiation therapy (SBRT): A planning study. Radiation Oncology. 2009;**4**(1):22

[27] de Moraes FY, Taunk NK, Laufer I, Neves-Junior WFP, Hanna SA, Carvalho HA, et al. Spine radiosurgery for the local treatment of spine metastases: Intensity-modulated radiotherapy, image guidance, clinical aspects and future directions. Clinics. 2016;**71**(2):101-109

[28] Taylor A, Powell M. Intensity-modulated radiotherapy—what is it? Cancer Imaging. 2004;**4**(2):68

[29] Shepard DM, Olivera G, Angelos L, Sauer O, Reckwerdt P, Mackie TR. A simple model for examining issues in

radiotherapy optimization. Medical Physics. 1999;**26**(7):1212-1221

[30] Cedric XY, Tang G. Intensity-modulated arc therapy: Principles, technologies and clinical implementation. Physics in Medicine and Biology. 2011;**56**(5):R31

[31] Teoh M, Clark C, Wood K, Whitaker S, Nisbet A. Volumetric modulated arc therapy: A review of current literature and clinical use in practice. The British Journal of Radiology. 2011;**84**(1007):967-996

[32] Salter BJ, Sarkar V, Wang B, Shukla H, Szegedi M, Rassiah-Szegedi P. Rotational IMRT delivery using a digital linear accelerator in very high dose rate 'burst mode'. Physics in Medicine and Biology. 2011;**56**(7):1931

[33] Bewes J, Suchowerska N, Jackson M, Zhang M, McKenzie D. The radiobiological effect of intra-fraction dose-rate modulation in intensity modulated radiation therapy (IMRT). Physics in Medicine and Biology. 2008;**53**(13):3567

[34] Moiseenko V, Duzenli C, Durand RE. In vitro study of cell survival following dynamic MLC intensity-modulated radiation therapy dose deliverya. Medical Physics. 2007;**34**(4):1514-1520

[35] Galvin JM, Ezzell G, Eisbrauch A, Yu C, Butler B, Xiao Y, et al. Implementing IMRT in clinical practice: A joint document of the American society for therapeutic radiology and oncology and the American association of physicists in medicine. International Journal of Radiation Oncology, Biology, Physics. 2004;**58**(5):1616-1634

[36] Rao M, Yang W, Chen F, Sheng K, Ye J, Mehta V, et al. Comparison of Elekta VMAT with helical tomotherapy and fixed field IMRT: Plan quality, delivery efficiency and accuracy. Medical Physics. 2010;**37**(3):1350-1359

[37] Langen KM, Papanikolaou N, Balog J, Crilly R, Followill D, Goddu SM, et al. QA for helical tomotherapy: Report of the AAPM task group 148. Medical Physics. 2010;**37**(9):4817-4853

[38] Chung Y, Yoon HI, Kim JH, Nam KC, Koom WS. Is helical tomotherapy accurate and safe enough for spine stereotactic body radiotherapy? Journal of Cancer Research and Clinical Oncology. 2013;**139**(2):243-248

[39] Kron T, Eyles D, Schreiner LJ, Battista J. Magnetic resonance imaging for adaptive cobalt tomotherapy: A proposal. Journal of Medical Physics. 2006;**31**(4):242-254

[40] Cashmore J. The characterization of unflattened photon beams from a 6 MV linear accelerator. Physics in Medicine and Biology. 2008;**53**(7):1933

[41] Titt U, Vassiliev O, Pönisch F, Dong L, Liu H, Mohan R. A flattening filter free photon treatment concept evaluation with Monte Carlo. Medical Physics. 2006;**33**(6Part1):1595-1602

[42] Jeraj R, Mackie TR, Balog J, Olivera G, Pearson D, Kapatoes J, et al. Radiation characteristics of helical tomotherapy. Medical Physics. 2004;**31**(2):396-404

[43] Antypas C, Pantelis E. Performance evaluation of a CyberKnife® G4 image-guided robotic stereotactic radiosurgery system. Physics in Medicine and Biology. 2008;**53**(17):4697

[44] Echner G, Kilby W, Lee M, Earnst E, Sayeh S, Schlaefer A, et al. The design, physical properties and clinical utility of an iris collimator for robotic radiosurgery. Physics in Medicine and Biology. 2009;**54**(18):5359

[45] van de Water S, Hoogeman MS, Breedveld S, Nuyttens JJ, Schaart DR, Heijmen BJ. Variable circular collimator in robotic radiosurgery: A time-efficient alternative to a mini-multileaf collimator? International Journal of Radiation Oncology, Biology, Physics. 2011;**81**(3):863-870

[46] Murai T, Hattori Y, Sugie C, Iwata H, Iwabuchi M, Shibamoto Y. Comparison of multileaf collimator and conventional circular collimator systems in Cyberknife stereotactic radiotherapy. Journal of Radiation Research. 2017;**58**(5):693-700

[47] Zach L, Tsvang L, Alezra D, Ben Ayun M, Harel R. Volumetric modulated arc therapy for spine radiosurgery: Superior treatment planning and delivery compared to static beam intensity modulated radiotherapy. BioMed Research International. 2016;**2016**:6

[48] Choi YE, Kwak J, Song SY, Choi EK, Do Ahn S, Cho B. Direct plan comparison of RapidArc and CyberKnife for spine stereotactic body radiation therapy. Journal of the Korean Physical Society. 2015;**67**(1):116-122

[49] Sahgal A, Chuang C, Hossain S, Petti P, Larson DA, Shrieve DC, et al. Comparisons of Novalis and CyberKnife® spinal stereotactic body radiotherapy treatment planning based on physical and biological modeling parameters. In: Radiosurgery. Switzerland: Karger Publishers; 2010;7:366-377

[50] Gallo JJ, Kaufman I, Powell R, Pandya S, Somnay A, Bossenberger T, et al. Single-fraction spine SBRT end-to-end testing on TomoTherapy, Vero, TrueBeam, and CyberKnife treatment platforms using a novel anthropomorphic phantom. Journal of Applied Clinical Medical Physics. 2015;**16**(1):170-182

[51] Zou KH, Warfield SK, Bharatha A, Tempany CM, Kaus MR, Haker SJ, et al. Statistical validation of image segmentation quality based on a spatial overlap index1: Scientific reports. Academic Radiology. 2004;**11**(2):178-189

[52] Krishnan J, Shetty J, Rao S, Hegde S, Shambhavi C. Comparison of rapid arc and intensity-modulated radiotherapy plans using unified dosimetry index and the impact of conformity index on unified dosimetry index evaluation. Journal of Medical Physics. 2017;**42**(1):14

[53] Bezjak A, Papiez L, Bradley J, Gore E, Gaspar L, Kong MF-MSP, et al. RTOG 0813 Seamless Phase I/II Study Of Stereotactic Lung Radiotherapy (SBRT) for Early Stage, Centrally Located, Non-Small Cell Lung Cancer (NSCLC) In Medically Inoperable Patients. 2009. Available from: https://www.rtog.org/ClinicalTrials/ProtocolTable/StudyDetails.aspx?study=0813

Section 3

Radioactivity

Chapter 8

Monitoring of Natural Radioactivity in Drinking Water and Food with Emphasis on Alpha-Emitting Radionuclides

Markus Zehringer

Abstract

Alpha spectrometry is an indispensable technique in the radiology lab for the analysis of natural radionuclides. While the powerful ICP/MS is used more and more for the analysis of uranium and thorium, other radionuclides, such as ^{226}Ra, are difficult to analyze with this technique due to their very high specific activities. The following chapter is introduced by a description of the problems, which may occur when working in the ultra-trace level. A description of the commonly used extraction and enrichment techniques for alpha nuclides and a short survey of the commonly applied detection techniques are given. The main application of alpha spectrometry in our laboratory is the monitoring of tap and mineral waters. Besides water, some specific food categories, such as fish, seafood, spices or healing earths, are monitored for their content of natural radionuclides.

Keywords: alpha spectrometry, mineral water, seafood, spices, healing earth

1. Introduction

Uranium and thorium are radionuclides with long half-lives. Their abundances in the earth's crust are 12–13 ppm rsp. 2.5 ppm for thorium and uranium. They exist in minerals, such as pitch blend (uranium) or monazites (thorium).

From the earth's crust, these elements migrate to the groundwater layers. There, they can be transferred to surface waters. Therefore, the consummation of water is a main direct source for the ingestion of alpha nuclides. In addition, they are taken up by aquatic organisms (fish, mussels) and enriched in these organisms (e.g. Po-210 (^{210}Po) is enriched in the intestinal tract of mussels and fish). Alpha nuclides are enriched in farmland soils by irrigation and the application of phosphate fertilizers, which contain respectable amounts of uranium and thorium. The consummation of vegetables cultivated on such farmlands and the consummation of meat and meat products from these farms is another possibility for the intake of alpha nuclides.

There exist three main decay series of natural radionuclides starting from the radionuclides uranium-238 (^{238}U), thorium-232 (^{232}Th) and uranium-235 (^{235}U). The ^{238}U decay chain produces several radionuclides, which are of radiological concern. These are ^{238}U, ^{234}U, radium-226 (^{226}Ra), radon-222 (^{222}Rn), lead-210

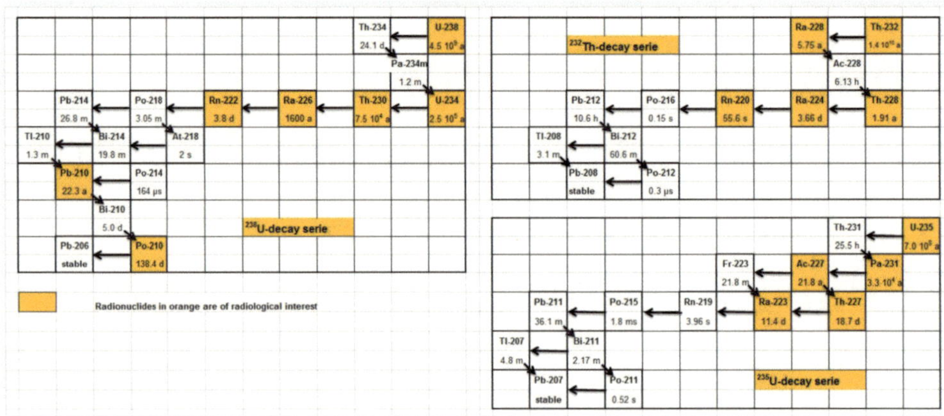

Figure 1.
Natural decay series of uranium and thorium. Alpha decays are symbolized with a horizontal arrow. The small diagonal arrows indicate a beta decay. In the nuclide box, the half-life is given. All data from [1].

(^{210}Pb) and ^{210}Po. The chain ends at the stable lead isotope ^{206}Pb. The ^{232}Th-decay chain comprises ^{232}Th, radium-228 (^{228}Ra), thorium-228 (^{228}Th), radium-224 (^{224}Ra) and radon-220 (^{220}Rn), which are of radiological relevance. The chain stops at the stable lead isotope ^{208}Pb. The Uranium-235 decay chain produces the following relevant radionuclides: ^{235}U, protactinium-231 (^{231}Pa), actinium-227 (^{227}Ac), thorium-227 (^{227}Th) and radium-223 (^{223}Ra). The chain stops at the stable lead isotope lead-207 (^{207}Pb). **Figure 1** shows the decay paths of the three decay series.

Radionuclides with longer half-lives, more than days, are of interest. Nevertheless, all radionuclides in the three decay chains are of radiological concern. The short-lived radionuclides cannot be considered separately. Therefore, the long-lived radionuclides and their short-lived daughter nuclides are assessed together. The conversion factor of a specific radionuclide considers the radiation of the following short-lived daughter nuclides.

Radon nuclides are radionuclides of the decay chains, which are gaseous. Therefore, they can be translocated by outgassing from the soil. That means, the radon nuclide and its follower nuclides are transported away from the soil or other environmental matrices that contain radium isotopes. Consequently, radon nuclides can be dislocated from its point of origin. ^{222}Rn can leave the soil due to its relatively long half-life of 3.8 days. One estimates that about 50% of the radon can leave the soil from a depth of 1 m [2]. In the air, the decay of the radon produces after several decay steps the longer-lived ^{210}Pb, ^{210}Bi and ^{210}Po, which are transported on dust particles through the atmosphere and may reach again farmland and surface waters. For the two other decay chains, this effect can be neglected. ^{219}Rn and ^{220}Rn are disintegrated before they have left the soil. In addition, the decay of these two radon species produces only short-lived radionuclides.

2. EU and Swiss legislation for radionuclides

Since 2018, in Swiss legislation most limit values for radionuclides for food were deleted after the adaption of the ordinances to the legislation of the European Union. The *Federal Ordinance on Contaminants and Constituents in Food* from 1994 was expired [3], and a new *Ordinance on Maximum Limits for Contaminants* was put into force in May 1, 2017 [4]. In this new ordinance, limit values for radionuclides in food at so-called emergency situations are included. For "planed exposition situations" (this means normal situations without any fallout event), no limit values at all are defined. Fortunately, limit values for radiocesium, an important beta-emitter in fallout, are regulated in the *Ordinance on the Importation and the Placing on the Market of Food, which is contaminated following the Accident of the Nuclear Power Station of Chernobyl (Chernobyl Ordinance)* [5] and the *Ordinance on Food originating in or consigned from Japan (Fukushima Ordinance)* [6]. Unfortunately, the limit values for radiocesium are different in each ordinance.

In Switzerland, a better legal basis exists only for drinking water. The *Ordinance on Drinking Water and Water from Public Baths and Shower Facilities (TBDV)* includes a limit value for uranium and guide values for tritium and radon and the parameter *indicative dose* [7]. This ordinance is based on the *European Council Directive 2013/51/Euratom* [8]. Fortunately, this council directive prescribes some

Radionuclide		**TBDV**	**Derived activity limits according to Euratom**	**WHO guidance level**
natural radionuclides				
Lead-210 (^{210}Pb)	Bq/L		0.2	0.1
Polonium-210 (^{210}Po)	Bq/L		0.1	0.1
Radium-226 (^{226}Ra)	Bq/L		0.5	1
Radium-228 (^{228}Ra)	Bq/L		0.2	0.1
Radon (^{222}Rn)	Bq/L	100	100	—
Thorium-228 (^{228}Th)	Bq/L			1
Thorium-230 (^{230}Th)	Bq/L			1
Thorium-232 (^{232}Th)	Bq/L			1
Uranium-238 (^{238}U)	Bq/L		3.0	10
Uranium-234 (^{234}U)	Bq/L		2.8	1
Uranium*	µg/L	30	—	
Artificial radionuclides				
Americium-241 (^{241}Am)	Bq/L	0.1	0.7	1
Carbon-14 (^{14}C)	Bq/L		240	100
Cesium-134 (^{134}Cs)	Bq/L		7.2	10
Cesium-137 (^{137}Cs)	Bq/L		11	10
Cobalt-60 (^{60}Co)	Bq/L		40	100
Iodine-131 (^{131}I)	Bq/L	6.2	6.2	10
Plutonium $^{239+240}$Pu	Bq/L		0.6	1
Strontium-90 (^{90}Sr)	Bq/L	4.9	4.9	10
Tritium (^{3}H)	Bq/L	100	100	10,000
Indicative dose (ID)	mSv	0.1	0.1	

TBDV, Swiss ordinance on drinking water and water for public baths and shower facilities [7]; Euratom, council directive 2013/51/Euratom [8]; WHO [9].
uranium is calculated from the activity of ^{238}U.

Table 1.
Guidance levels for drinking water in EU and Switzerland compared to the WHO guidance levels.

more guidance limits, “derived activity limits”, for artificial and natural radionuclides. Therefore, these activity limits are also applicable in Switzerland. **Table 1** summarizes the actual valid guidance and limit values for drinking water existing in Europe and Switzerland.

3. Collection and conservation of water samples

In this short chapter, important tips for a good analytical practice are given. Sample collection and storage can be the source of basic errors, which can no more be eliminated, even using a sophisticated analytical technique. At first, we must remember the concentration level we are working at.

3.1 Specific activity

When analyzing radio traces, one must keep in mind the very low chemical concentrations one is dealing with. Due to short half-lives (between hours and days), the specific activity of many species is very high. This means that very low, mostly non-weighable chemical concentrations correspond to measurable activities in the mBq range. Practical work with such traces requires the addition of inactive carriers. The following table illustrates this effect [10]. ^{232}Th and ^{238}U are exceptions. Both nuclides decay with half-lives of billion years. Therefore, common activities correspond to weighable chemical amounts (**Table 2**).

These very low concentrations must be kept in mind, when samples are collected and prepared for radio trace analyses. The problems, which can occur, are losses of the analytes or contaminations during collection, transport, conservation and preparation of samples.

3.2 Sample collection and storage

From natural waters, it is a major problem to obtain representative samples, e.g. out of a river. Trace element concentrations depend on depth, salinity and turbidity of a river or lake. Groundwater samples collected by using a pump should be taken some minutes after beginning of the pumping to avoid the collection of the standing water in the pipe. At this point, we cannot explicate profoundly the challenges for adequate sampling. We recommend consulting the relevant literature, e.g. [11–13].

Radionuclide	Half-life	Specific activity MBq/kg	Activity limits in tap water (Bq/L)*	Corresponding concentration µg/L (ppb)
^{222}Rn	3.8 d	5.7 10^{12}	100	2 10^{-11}
^{210}Po	138 d	1.66 10^{11}	0.10	6 10^{-10}
^{210}Pb	22.3 a	2.8 10^{9}	0.20	4 10^{-8}
^{228}Ra	5.75 a	1.0 10^{10}	0.20	1 10^{-7}
^{226}Ra	1600 a	3.4 10^{7}	0.50	3 10^{-6}
^{238}U	4.4 10^{9} a	12.5	3	8.1
^{232}Th	1.4 10^{10} a	4.06	1	24

**Activity limits according to ordinance on drinking water and water for public baths and shower facilities (TBDV) [7].*

Table 2.
Specific activities of some dose-relevant natural radionuclides.

Another important point is the choice of adequate sample containers. For metal trace analysis, no glass containers should be used. The active surface of glass bottles acts like an ion exchanger. SiOH groups are potent ligands for metal ions and may adsorb ions from the water sample. One possibility to omit these adsorptive effects is the conservation of the sample with mineral acid, such as nitric acid, hydrochloric acid or sulphuric acid. SiOH groups are then protonated, and the adsorption of cations is hindered. One can also add carrier ions to obtain the same effect. This is a commonly used technique in radiochemical analyses. Addition of stable isotopes of the analytes in higher concentrations prevents such adsorptive effects. Another possibility is to precipitate radio traces by adding carriers (co-precipitation). Another possibility is to catch and stabilize the analyte with a chelating agent, such as EDTA, to prevent absorptive losses. Samples may also be dried at 105°C for conservation or by freeze-drying, for example, of milk.

To overcome the possibility of such losses of analytes, we recommend using plastic bottles of polyethylene, PVC or Teflon, whenever possible. Pyrex (borosilicate glass) or soda glass containers should be avoided [13, 15, 16]. The containers should be soaked in diluted acid and rinsed thoroughly with distilled water before use. Before the sampling, the bottles are rinsed several times with the sample water [12].

It is also advisable to stabilize samples with the addition of conc. Nitric or hydrochloric acid, if the applied procedure does allow this (pH should be below 1). A stabilization is necessary when samples cannot be analyzed immediately (losses or transformation by bacteria, co-precipitation with suspended matter, etc. must be avoided) [11, 12, 16].

A special case is the sampling of the gaseous radon. Here, losses may occur when samples are collected under turbulences or bottles are not completely filled, leaving some headspace. Radon will outgas partially. Here, it is advisable to collect the water without turbulences in glass bottles under water, if possible. The bottles must be filled to the top without letting back any air bubbles. For sampling of water from taps or valves, a simple method is to let stream the water through a plastic tube connected to a funnel and to fill the bottle from the bottom up to the top without turbulences. Then, the bottle is closed with a glass stopper displacing the water in the bottleneck. Normally, the radon activities in water are in the Bq/L range, so loss effects by adsorption effects on glass walls may not be noticed. The bottles should be transported to the lab at low temperature, but a freezing of the sample should be avoided. The samples must be analyzed within a week due to the fast disintegration of the radon [11].

3.3 Quality of standards and chemicals

One should take special care to choose acids and other chemicals of high quality. They should be of sufficiently high purity. Concentrated mineral acid solutions (e.g. 65% nitric acid, 95% sulphuric acid) and other chemicals should be of the same quality used for ICP, AAS or XRF analysis. These are, for example, mineral acids of suprapur quality from Sigma-Aldrich, Merck, Roth, etc. For these products, specification data sheets with the declaration of minimum trace amounts of most metals are available. For every method, a careful check of the whole blank (chemicals used, demineralized water, glassware for sample preparation, etc.) is important and is a matter of course.

4. Sample preparation techniques

Alpha rays have a short reach due to its high mass and dual positive mass (two protons and two neutrons). Therefore, the matrix absorbs most alpha radiation. Alpha rays can leave the sample only from very thin surfaces. To detect alpha rays, one must eliminate the matrix without losing the alpha nuclides. The following survey is not a complete review of the commonly used techniques. It is more focused on the applied procedures at the state laboratory of Basel City. This focus allows us to report from our long-time experience in alpha spectrometry.

4.1 Oxidation of the matrix

There are several common techniques for the preparation of samples. For non-volatile and thermal stable analytes, solid samples can be calcinated in an oven (dry ashing). For volatile analytes, e.g. ^{210}Po, where losses are possible at temperatures over 200°C or radiocesium (400°C), the matrix can be oxidized with mineral acid/ peroxide in a microwave oven at moderate temperatures around 200°C.

In water samples, the analytes can be concentrated by evaporation or distillation of the water. Other procedures include direct evaporation of the sample on surfaces or by vacuum sublimation (e.g. Frisch-grid ionization chambers or gas proportional counters). The German DIN prescribes the evaporation of the water phase in the presence of barium as a carrier for the analysis of ^{226}Ra. After a second precipitation with sulphate as a cleanup to remove thorium and polonium, the analyte can be measured [17].

4.2 Liquid-liquid extraction

Liquid samples can be water samples or aqueous extraction solutions of solid samples. They may be extracted with nuclide-specific, extractive cocktails. Jack McDowell et al. have developed a set of extractants/fluors for different alpha nuclides [18]. Some special applications with the photon electron rejection alpha liquid spectrometry (PERALS) system are described in [19–21]. For the extraction of uranium with the extractor/scintillator URAEX, several investigations to optimize analytical procedures were published [22–24]. Véronneau et al. [25] investigated the extraction of polonium with different extractors/scintillators. Our preferred system is PPBO[1]/TOPO[2], commercially available as POLEX™. The PERALS system combined with a set of specific extractant/scintillator is a powerful analytical tool for the analysis of alpha nuclides (**Table 3**) [27].

Reagent	Extractable radionuclides	Literature data
ALPHAEX	U, Th, Pa, Hf, Zr, Pu (IV)	23, 28
POLEX	Specific for ^{210}Po, ^{237}Np	25, 29, 30
RADONS	Specific for ^{222}Rn	31
RADAEX	^{226}Ra (and daughters)	20, 21, 29, 32
STRONEX	Specific for ^{90}Sr	33
THOREX	^{228}Th, ^{230}Th, ^{232}Th and Zr, Hf, U, Eu, In and other nuclides	34, 35
URAEX	Specific for ^{234}U and ^{238}U	21, 23, 24

Table 3.
Extractant/fluor solutions for PERALS-α-spectrometry [26].

[1] 2-(4′-biphenylyl)-6-phenylbenzoxazole
[2] Trioctylphosphine oxide

Many other extraction systems were published. For example, Leeuwen et al. published an overview on the selective extraction of radium with means of chelators [36, 37].

4.3 Adsorptive surfaces

The use of some specific active surfaces, which are suitable for the adsorption of radionuclides from aqueous solutions and are commercially available, was published.

Such selective phases can be made of pure metals, such as copper, nickel or silver. ^{210}Po will auto adsorb on such a surface in an acidic, reductive milieu using ascorbic acid or hydroxylamine [38–41]. For recovery control of the adsorption process, polonium-208 (^{208}Po) or polonium-209 (^{209}Po) can be added as tracers.

Manganese dioxide-coated surfaces adsorb radium nuclides selectively. Based on the work of Moore and Reid [42], Glöbel and Berlich [43] and Surbeck [44], robust procedures for the analysis of radium in water were developed by Eikenberg et al. and Surbeck [45, 46]. Ra disks are commercially available from nucfilm [47].

For uranium, there also exists a selective surface based on diphonix resin supported on polycarbonate [48]. Diphonix-coated disks are commercially available from nucfilm [47]. They work well below activities of 1 Bq/L because of the restricted load capacity. For example, Surbeck suggests analyzing for uranium only after the elimination of radium by adsorption with a MnO_2 disk. Otherwise overloading effects with radiumnuclides may be noticed [49].

Alpha nuclides in a liquid milieu can be adsorbed electrolytically onto stainless steel disks (electroplating, electrodeposition). The *Mitchell method* describes the deposition of actinides (uranium, americium, polonium) in a HCl milieu [50]. The *Talvitie method* is used for the deposition of actinides (uranium, thorium) in the presence of a HSO_4^-/SO_4^{2-} milieu [51]. Many applications were published for uranium, thorium, radium, plutonium and other alpha nuclides [52, 53]. In our laboratory we have experiences in analyzing uranium and thorium in acid extract solutions of spices [54]. Our method is based on working sheets from the Paul Scherrer Institute (PSI), which describes the production of a homemade electrodeposition unit and the application for the enrichment of uranium or thorium on stainless steel planchettes [51, 55–58]. Frindik et al. give a method for the determination of many alpha nuclides, including such as plutonium and americium [59].

4.3.1 Micro-precipitation

For radium, micro-precipitation processes based on manganese dioxide are well used. Many alpha nuclides, such as americium or plutonium, are analyzable. To equalize losses of analytes in the precipitation step, the use of radioactive tracers (internal standards) is important [60]. Thorium, uranium, plutonium, americium and curium species are analyzable [61, 62].

4.3.2 Filtration

Radionuclides are collected on impregnated filters more or less selectively. Best examples are radium-specific filters from 3 M Empore (Radium Rad disks). ^{228}Ra can be analyzed indirectly via its short-lived daughter ^{228}Ac, which is built on the filter surface by the disintegration of the enriched ^{228}Ra. This is a very interesting approach for the analysis of the dose-relevant ^{228}Ra [63, 64].

5. Measurement equipment

For a general survey, we recommend the *Handbook of Radioactivity Analyses* [65]. Another somewhat older standard book is *Radiation Detection and Measurement* from Knoll [66]. In this chapter, we describe the most used analytical techniques for alpha nuclides as we know.

5.1 Alpha particle spectrometry

Passivated implanted planar silicon detectors (PIPS) or silicon barrier detectors are widely used in analyzing alpha nuclides. It is necessary to obtain thin, homogenous alpha sources. Thick sources show broad peaks due to some degradation cause by self-adsorption. High-resolution alpha spectrometry needs thin sources. For this purpose, stainless steel plates or adsorptive surfaces are ideal. Very thin micro precipitates are also suitable. Most alpha nuclides are energy resolved, with one exception. The alpha energies of plutonium-239 (^{239}Pu) and plutonium-240 (^{240}Pu) differ only 10 keV. They may not be resolved in the alpha spectrum and therefore often given as the activity sum of both nuclides.

Detector systems are available from Ortec [67] and Mirion (former Canberra) [68]. They are available as multichamber systems to handle the long counting times (e.g. 24 hours) in the daily routine. In our laboratory, we use PIPS detectors to detect alpha nuclides of polonium, radium, uranium and thorium (**Figure 2**).

5.2 Liquid alpha scintillation techniques

Liquid scintillation techniques (LSC) with an alpha/beta discrimination possibility are often used [69]. An example is the analyses of radon. Very sensitive analyses of radon in water samples are possible, when the large beta background of the radon daughters ^{214}Bi and ^{214}Pb is discriminated. The consecutive alpha decays of ^{222}Rn, ^{218}Po and ^{214}Po are counted (cross alpha counting) [70]. Three main procedures for the analysis of radon with LSC are possible. The "direct" method consists in mixing of the water with a water immiscible cocktail. The mixture is then counted with LSC. In the indirect method, radon is transferred from the water into the cocktail phase and then analyzed. The third method uses a purging of the radon into a cocktail phase. The simplest and most precise method is the direct method, because only one step in the sample preparation is necessary. Low detection limits are achievable with a small effort in sample preparation (detection limit of about 0.2 Bq/L).

A special alpha scintillation technique is commercially available from ORDELA [26]. The PERALS called alpha spectrometric technique is based on the following procedure. The analyte is extracted with a selective extracting agent, which contains a specific fluor. With the help of the fluor, α-, β- and γ-rays are converted into photons. The β- and γ-photons of the extract solution are discriminated by the use of a pulse shape discriminator. The discrimination is based on the longer relaxation times of alpha decays. The discriminator eliminates the fast-relaxing β- and γ-induced photons. As resolution of the emissions is much poorer than PIPS detectors, nuclide-specific extractions systems must be used. Methods are available for the analysis of polonium, radon, radium, uranium, thorium and other alpha nuclides [18], see 4.2 (**Figure 2**).

5.3 Gas proportional counting

Gas proportional counting is a suitable analyzing technique when equipped with the possibility to separate alpha from beta rays. Loaded planchettes or disks and thin

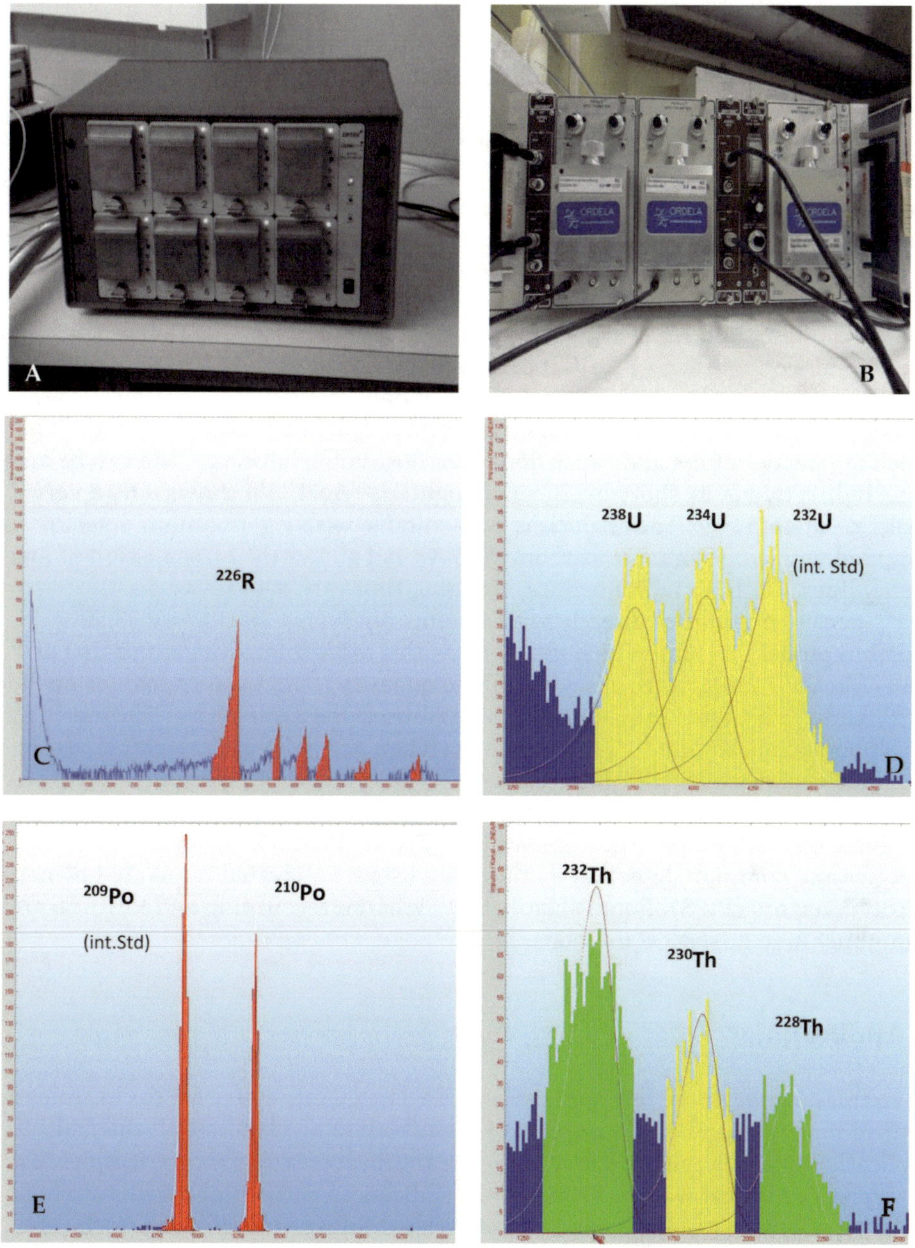

Figure 2.
Common equipment for alpha spectrometry. A, PIPS alpha spectrometer with eight counting cells; B, PERALS alpha liquid scintillation counters; C, alpha spectrum of radium in mineral water; D, PERALS spectrum of uranium in san Pellegrino mineral water; E, PIPS alpha spectrum of polonium in a mineral water; F, PERALS spectrum of thorium in a mineral water.

precipitates can be counted. Methods exist for polonium and other alpha emitters. It is also possible to extract and analyze alpha and beta nuclides, e.g. ^{210}Pb and ^{210}Po.

5.4 Track detectors

Track detectors are used since many years for the measurement of ^{222}Rn in air. The tracks on an exposed foil produced by alpha particles are counted with a microscope after chemical etching of the foil.

5.5 Gamma-ray spectrometry

Gamma-ray spectrometry is only suitable for the analyses of alpha nuclides in samples, where the analytes are present in higher activities (soil, sediments, etc.). Uranium, for example, is only detectable via its daughter nuclides (e.g. ^{234}Th for ^{238}U). Therefore, it is important to render data plausible according to the individual decay time within the decay chains. Radionuclides, which are in secular equilibrium, should show equal activities. Interferences may pretend excessive activities.

A second challenge is the weak photon emissions of alpha nuclides. In addition, these emissions are in the lower keV range, and the correlation to a specific radionuclide is often doubtful because of interferences. ^{232}Th is such a radionuclide with a main emission line at 63.8 keV. It can easily be mixed with ^{234}Th (63.3 keV). Such lines of weak intensity need counting times of days to get reasonable signals [71]. In drinking water analysis, there are two important radionuclides, ^{228}Ra and ^{210}Pb, which are beta emitters and cause dominant dose contributions. ^{228}Ra can be analyzed indirectly via its short-lived beta daughter ^{228}Ac. ^{210}Pb shows only a very weak line at 46.5 keV. The analysis is only suitable with a germanium detector equipped with a beryllium or carbon window that allows the transmission of low-energy lines. For both radionuclides, counting times are days or weeks.

^{226}Ra can be analyzed directly with its emission line at 186.21 keV (3.56% emission probability) but which interferes with a more intensive gamma line of ^{235}U (185.72 keV, 57.2%). Here, it is advisable to quantify ^{226}Ra via its daughter nuclides (^{214}Bi and ^{214}Pb), after the radioactive equilibrium is established. ^{228}Th and ^{224}Ra are alpha nuclides of the ^{232}Th decay chain. Their activities are calculated via the daughter nuclides (^{212}Pb, ^{212}Bi). ^{238}U may be analyzed via its daughter ^{234}Th. As mentioned before, these radionuclides are only quantifiable in samples, which show relatively high activities (e.g. soil samples) [71]. In **Table 4**, radionuclide decay energies and common detection methods are listed. Passivated implanted planar silicon detectors (PIPS), liquid scintillation alpha/beta-counting and PERALS are commonly used analytical methods.

6. Applications

In this chapter, some few applications of alpha spectrometry for the examination of drinking water and food samples are described. In our laboratory, the main application of alpha spectrometry is for the routine control of water samples, e.g. tap water or mineral waters. Fish and seafood, spices and healing earths are food categories, which may have incorporated relevant quantities of alpha nuclides.

6.1 Monitoring of mineral waters from the Swiss market

Natural mineral water should be water of good microbiological quality. It is collected in groundwater layers or rock formations. This water reaches the surface by one or several naturally or artificially built exit points and can be collected there. Mineral waters are classified according to their mineral contents (from very low mineral content to high mineral content) or specified for a dominant constituent (e.g. rich in magnesium, iron, fluoride). Another classification respects the content of carbonic acid (non-carbonated water, sparkling water, etc.). In all cases, minimum or maximum concentrations are defined in the Swiss Ordinance on Beverages [72] and other ordinances. Radionuclide activities in mineral waters may vary in a wide range and are caused by complex solubility and transport processes in the aquifer. The hydrogeological conditions, the uranium and thorium content of the

Radionuclide	Half-life	Alpha energies (MeV), % branching	Common methods comments
^{210}Pb	22.3 y	beta	LSC Gas prop. Counter: indirect via ^{210}Bi Gamma spectrometry: weak line at 46.5 keV
^{210}Po	138.4 d	5.407,5 (100)	LSC: PERALS PIPS: Ag and Cu disk
^{231}Pa	3.28 10^4 y	4.762,5 (1,8) 4.795,4 (1,2) 4.819,8 (8,4) 4.939,1 (1,4) 5.023,0 (2,9)	PIPS: electrodeposition Gamma spectrometry
^{224}Ra	3.66 d	5.547,9 (5,3) 5.788,9 (94,7)	LSC: PERALS PIPS: electrodeposition PIPS: MnO_2 disk Gamma spectrometry
^{226}Ra	1600 y	4.684,0 (5,95) 4.870,5 (94,0)	LSC: PERALS PIPS: electrodeposition PIPS: MnO_2 disk Gamma spectrometry: interfering line at 86 keV
^{228}Ra	5.75 y	beta	LSC Gamma spectrometry: via ^{228}Ac
^{220}Rn	55.6 s	6.404,7 (100)	PIPS: PERALS
^{222}Rn	3.83 d	5.590,3 (100)	LSC: several specific methods
^{228}Th	1.91 y	5.520 (73,4) 5.436 (26,0)	LSC: PERALS B: electrodeposition
^{230}Th	7.54 10^4 y	4.620,5 (23,4) 4.687,0 (76,3)	LSC: PERALS PIPS: electrodeposition
^{232}Th	1.41 10^{10} y	4017,8 (21,0) 4081,6 (78,9)	LSC: PERALS PIPS: electrodeposition
^{234}U	2.45 10^5 y	4.804,5 (28,4) 4.857,6 (71,4)	LSC: PERALS and others PIPS Gamma spectrometry (indirect)
^{235}U	7.04 10^8 y	4.441,7 (18,8) 4.474,0 (7,25)	LSC: PERALS and others Gamma spectrometry
^{238}U	4.47 10^9 y	4.220,2 (22,3) 4.269,7 (77,5)	LSC: PERALS and others PIPS Gamma spectrometry (indirect)
^{239}Pu	2.41 10^4 y	5.192,8 (11,9) 5.231,5 (17,1) 5.244,5 (70,8)	LSC: PERALS PIPS
^{240}Pu	6561 y	5.210,5 (27,2) 5.255,8 (72,7)	PIPS
^{241}Am	432.4 y	5.479,3 (1,7) 5.534,9 (13,2) 5.578,3 (84,5)	LSC: PERALS PIPS Gamma spectrometry
^{244}Cm	18.1 y	5.858,9 (23,3) 5.901,7 (76,7)	LSC: PERALS PIPS

Taken from: [1].

Table 4.
Radionuclide decay energies and common detection methods.

rock formations, solubility behaviour and other chemical characteristics of the radionuclides are crucial for their presence in the water phase.

Besides tap water, mineral waters are the most consumed beverage in Switzerland. Nearly one billion liters are consumed yearly. Today, about 57% of the mineral waters are from Swiss production. 43% are originated from European countries, such as Italy, France, Germany and others [73]. The consummation of such high quantities of water requires rigorous monitoring. However, the law prescribes the declaration of the mineral contents only. Besides a microbiological survey, the focus is on contaminants from agriculture, industry and other productions. In past years, contaminations were detected sporadically. For example, traces of benzene in a French mineral water, *E. coli* bacteria in a mineral water infiltrated by contaminated seawater or high amounts of radium or uranium in French and Portuguese products gave reason for small scandals and violations of the law. Increased uranium activities in groundwater, originating from the intensive use of phosphate fertilizers, gave reason for the establishing of a monitoring programme for drinking water plants in Germany.

Natural and artificial radio contaminants were first monitored systematically in the USA and Germany [74].[3] In Switzerland, first investigations of mineral waters are from 1990. The Federal Office of Public Health mandated the Swiss Paul Scherrer Institute to analyze mineral waters available from the Swiss market [75]. In 2006 and 2007, the state laboratory of Basel City analyzed mineral waters in collaboration with the Office for food safety and veterinary affairs of the state of Basel-Country with the focus on uranium, tritium, radon and heavy metals [76, 77].

In 2018, our laboratory analyzed 46 mineral waters from the Swiss market. It was a complete investigation of natural and artificial radionuclides according to the TBDV [7]. Since 2018, the food law has changed. The former Ordinance of Food Contaminants and Constituents (FIV) was invalidated [3]. Therefore, no more limits exist for radionuclides in mineral waters. Consequently, it was necessary to validate the results according to the guide and limit values of the Swiss drinking water ordinance (**Table 1**).

Mineral water samples were collected at stores of the city of Basel. 21 products were Swiss mineral waters, followed by 12 mineral waters from Italy, 3 from Serbia and 2 each from Germany and France. Other countries were Fiji islands, Spain, Portugal, Norway, Kosovo and Croatia. The samples were analyzed by using the following sample preparation/measurement scheme (**Table 5**).

The aim of such investigations is always to estimate internal doses by consumption of the analyzed food category. Therefore, all dose-relevant radionuclides must be analyzed. These are natural radionuclides and anthropogenic radionuclides, such as ^{3}H, ^{40}K or ^{241}Am. According to TBDV, the total dose is the sum of the individual doses of all natural radionuclides, with the exception of potassium-40 (^{40}K), tritium (^{3}H) and ^{222}Rn with its short-lived daughters. So, uranium, radium, polonium, thorium and lead nuclides are the main nuclides which contribute to the dose. The doses due to the ingestion of individual radionuclides were calculated as follows:

$$D_i = c_i * e_{ing,i} * U, \quad (1)$$

with

---

[3] Federal ministry for the environmental conservation, construction and nuclear security (2017). Guidelines on compliance with the requirements of the Drinking Water Ordinance in the testing and evaluation of radioactive substances in drinking water. Drinking water is tested for its concentrations of radioactive substances nationwide.

Radionuclide	Sample preparation	Sample volume	Analytical technique	Quantification limit
^{234}U, ^{238}U	Extraction with 5 mL of URAEX	500 mL	PERALS-α-spectrometry	4 mBq/L
^{224}Ra, ^{226}Ra	Adsorption on MnO_2 disk	200 mL	PIPS-α-spectrometry	2 mBq/L
^{228}Ra	No sample preparation	1000 mL	γ-spectrometry (via ^{228}Ac)	50 mBq/L
^{210}Po	Adsorption on Ag disk	100 mL	PIPS-α-spectrometry	5 mBq/L
^{210}Pb	Adsorption on Ni disk	200 mL	b-gas proportional counter	50 mBq/L
^{228}Th, ^{230}Th, ^{232}Th	Extraction with 5 mL of THOREX	500 mL	PERALS-α-spectrometry	2 mBq/L
^{222}Rn	1:1 mix with Maxilight cocktail	10 mL	α-counting with LSC	0.4 Bq/L
^{3}H	Mix with 12 mL of Ultima Gold LLT cocktail	8 mL	β-counting with LSC	2 Bq/L
^{90}Sr	Extraction with 8 ml of STRONEX	1000 mL	β-counting with LSC	0.05 Bq/L
^{60}Co, ^{131}I, ^{134}Cs ^{137}Cs, ^{241}Am	No sample preparation	1000 mL	γ-spectrometry	0.05–0.1 Bq/L

Table 5.
Sample preparation and analyzing techniques used for the investigation of mineral waters at the state laboratory of Basel City.

c_i: Activity concentration of the radionuclide i
$e_{ing,i}$: Ingestion factor of radionuclide i [78]
U: Consummation rate (720 L for adult persons and year)

The sum of these individual doses is defined as *indicative dose*. The dose coefficients are listed in **Table 6**.

Results of the monitoring of mineral water in Switzerland in 2018 are presented in **Table 7**.

Uranium was detectable in 45 of 46 mineral waters. The mean activities were 0.02 ± 0.02 rsp. 0.02 ± 0.03 for ^{234}U and ^{238}U. The corresponding chemical concentration was 1.9 ± 2.0 μg U/L, fulfilling the limit value of 30 μg/L. The ratio of $^{238}U/^{234}U$ of 1.04 shows an undisturbed equilibrium between the two uranium nuclides. The Italian mineral waters *San Pellegrino* and *Varanina* contained 183 and 269 mBq/L of both uranium nuclides, corresponding to 7.0 rsp. 10.6 μg/L.

^{228}Ra is the dominant radionuclide of the radium species. Eleven samples showed activities of ^{228}Ra with a mean of 0.07 ± 0.05 Bq/L. ^{224}Ra and ^{226}Ra were only found in traces (mean 0.04 Bq/L for ^{226}Ra) with one exception. *Pedras Salgadas*, a mineral water from Portugal, showed an activity of 1.4 Bq/L of ^{226}Ra. Here, the guide value of 0.5 Bq/l was violated. Before 2017, this mineral water was banned from the Swiss market due to the then existing limit value of 1 Bq/L for liquid food. ^{224}Ra was only detected in two samples (<0.002–0.10 Bq/L).

^{210}Pb and ^{210}Po are built at the end of the decay chain of ^{238}U. ^{210}Pb has a relatively long half-life of 22.3 years and is the mother nuclide of ^{210}Bi and ^{210}Po. It is a beta nuclide and commits to the indicative dose. ^{210}Po is an alpha nuclide with a high alpha energy of 5.3 MeV and a half-life of 138 days. Therefore, it is also dose-relevant. Fourteen mineral water samples showed a measurable activity of ^{210}Pb.

Radionuclide	Infants (1–2 years)	Children > 10 years	Adult persons
Americium-241 (^{241}Am)	0.37	0.22	0.20
Lead-210 (^{210}Pb)	3.6	1.9	0.69
Cesium-134 (^{134}Cs)	0.016	0.014	0.019
Cesium-137 (^{137}Cs)	0.012	0.010	0.013
Iodine-131 (^{131}I)	0.180	0.052	0.022
Cobalt-60 (^{60}Co)	0.027	0.011	0.003
Polonium-210 (^{210}Po)	8.8	2.6	1.2
Radium-224 (^{224}Ra)	0.66	0.26	0.065
Radium-226 (^{226}Ra)	0.96	0.80	0.28
Radium-228 (^{228}Ra)	1.70	1.70	1.70
Radon (^{222}Rn)	0.02	0.02	0.01
Strontium-90 (^{90}Sr)	0.073	0.06	0.028
Thorium-228 (^{228}Th)	0.37	0.14	0.072
Thorium-230 (^{230}Th)	0.41	0.24	0.21
Thorium-232 (^{232}Th)	0.45	0.29	0.23
Tritium (^{3}H)	4.8E − 11	2.3E − 11	1.8E − 11
Strontium-90 (^{90}Sr)	0.073	0.06	0.028
Uranium-238 (^{238}U)	0.12	0.068	0.045
Uranium-234 (^{234}U)	0.13	0.074	0.049

Table 6.
Ingestion factors of the Swiss radiological protection ordinance [78] based on the ICRP concept. All data in μSv/Bq.

Radionuclide		Mean ± s.d.	Min	Max	n
^{234}U	Bq/L	0.02 ± 0.02	0.002	0.14	45
^{238}U	Bq/L	0.02 ± 0.03	0.002	0.13	45
^{nat}U	μg/L	1.9 ± 2.0	0.1	10.6	45
^{224}Ra	Bq/L	0.003 ± 0.02	0.002	0.11	2
^{226}Ra	Bq/L	0.04 ± 0.21	0.002	1.4	23
^{228}Ra	Bq/L	0.07 ± 0.05	0.05	0.44	11
^{222}Rn	Bq/L	1.3 ± 1.4	0.4	4.4	27
^{210}Po	Bq/L	0.04 ± 0.06	0.01	0.23	14
^{210}Pb	Bq/L	0.10 ± 0.06	0.05	0.27	14
^{228}Th	Bq/L	0.003 ± 0.002	0.002	0.01	22
^{230}Th	Bq/L	0.002 ± 0.002	0.002	0.01	12
^{232}Th	Bq/L	0.003 ± 0.004	0.002	0.02	20

s.d., standard deviation; n, number of positive samples from a total of 46 samples.

Table 7.
Results from the monitoring of mineral water in Switzerland.

DOI: http://dx.doi.org/10.5772/intechopen.90166

The mean was 0.10 ± 0.06 Bq/L. The *Pedras Salgadas* mineral water contained 0.3 Bq/L and therefore was over the guidance value of 0.2 Bq/L. Fourteen mineral water samples contained ^{210}Po in low activities (mean: 0.04 ± 0.06 Bq/L). In an Italian mineral water, the guidance limit of 0.1 Bq/L was overridden (0.23 Bq/L).

Thorium nuclides were found in low activities but in almost all mineral waters. The mean sum of the three nuclides was 0.01 Bq/L. These low activities are explained with the insolubility of thorium in water. Thorium species bind mainly on particles or co-precipitate with minerals and are therefore removed from the water phase.

^{222}Rn was found in 27 of the 46 samples with a low, mean activity of 1.3 ± 1.4 Bq/L (0.4–4.4 Bq/L). Most of the radon is lost during the production and transport of the mineral waters, except the Portuguese *Pedras Salgadas*, where radon is constantly produced by the disintegration of ^{226}Ra.

Artificial radionuclides, such as ^{3}H, ^{90}Sr and ^{137}Cs, were detectable in traces. Eight samples contained 0.01 ± 0.01 Bq/L ^{137}Cs. One sample showed traces of ^{90}Sr (0.05 Bq/L). ^{3}H was detectable in only two samples 1.2 ± 0.9 Bq/L. Radiocesium and radiostrontium are the main components of the global fallout and fallout from NPP accidents (Chernobyl). Their migration into the soil is a slow process. Therefore, disintegration may be fast enough to reduce the activities to amounts below the detection limits in the groundwater phase.

The calculated internal doses of the mineral waters (indicative doses) were 0.06 ± 0.60 mSv/a. Most samples fulfilled the guidance value of 0.1 mSv/a. The indicative dose of five mineral waters was over the guidance value. The highest dose was calculated for the Portuguese product *Pedras Salgadas.* We calculated a dose of 0.46 mSv/a for adult persons. Four other mineral waters showed doses between 0.12 and 0.27 mSv/a. Elevated activities of ^{226}Ra, ^{228}Ra and ^{210}Pb were the cause for these higher doses (**Figure 3**).

6.2 Analysis of healing earths

Siliceous earths are widely used in the food industry as a food supplement. They are deposits of the silica shells of diatoms (main constituent of marine phytoplankton). These layers are extracted in mines. Siliceous earths incorporate foreign atoms in the crystal lattice, such as radionuclides of the natural decay series

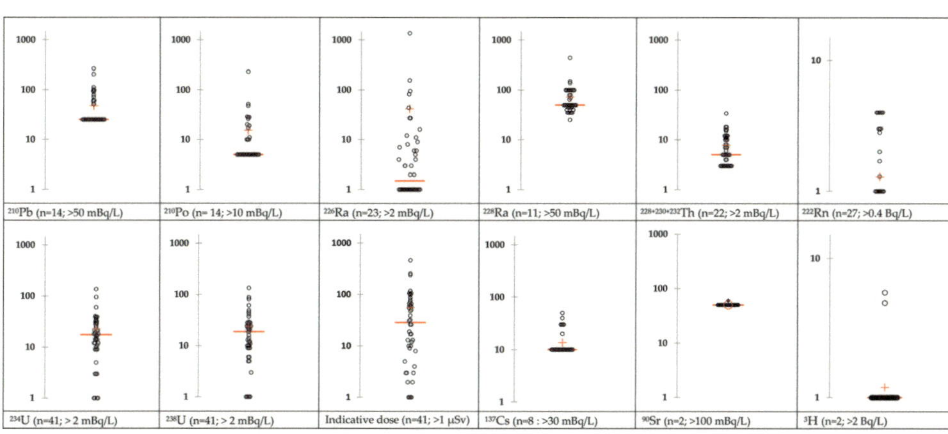

Figure 3.
Results from the investigation of mineral waters on the Swiss market 2018. All data in mBq/L, except indicative dose (µSv). n: Number of samples with activities over the detection limit (given as ">x mBq/L").

Radionuclide	Mean ± s.d.	Samples*	Min	Max
^{224}Ra	25.3 ± 21.5	30/32	1.7	63
^{226}Ra	42.1 ± 33.6	30/32	2.9	133
^{228}Ra	27.2 ± 23.5	30/32	2.0	67
^{234}U	56.9 ± 29.1	18/26	15.4	115
^{235}U	4.0 ± 2.1	07/26	1.5	6.9
^{238}U	75.4 ± 56.7	22/26	6.5	227
^{227}Th	2.6 ± 1.8	17/29	0.7	6.0
^{210}Pb	183 ± 152	6/24	49	428
^{210}Po	11.0 ± 9.0	22/24	2.0	42

s.d.: standard deviation.
**Number of positive samples/total of samples.*
All data in Bq/kg dry weight and from [78, 79].

Table 8.
Investigation of healing earths on the Swiss market.

of uranium and thorium. In 2008, the state laboratory of Basel City analyzed siliceous earth products on the Swiss market with α- and γ-spectrometry. In two products, the limit values for ^{226}Ra and ^{210}Po were exceeded (>50 Bq/kg). Furthermore, the annual dose by regular consummation of one product from California, USA, reached 0.5 mSv, half of the permitted yearly dose (1mSv). Consequently, this product was withdrawn from the Swiss market [79]. In 2010, a second inspection of the healing earths on the Swiss market showed that two products from German production slightly exceeded the limit values of ^{226}Ra and ^{228}Ra. The calculated, annual doses of these products when regularly consumed reached 0.1mSv/year [80]. Finally, we noted that healing earths can lead to doses up to 0.5 mSv. At last, materials based on silica for industrial use (e.g. as adsorbents or filter media in a chemical laboratories) can lead to the contamination of the environment when such materials are disposed or burnt (**Table 8**).

6.3 Radiological investigation of spices

Radionuclides from the uranium and thorium series may be enriched in plants and therefore also in spices. Until 2017, Swiss legislation included limit values for natural radionuclides in spices. For radionuclides of the group I (^{224}Ra, ^{228}Th, ^{234}U, ^{235}U, ^{238}U), a cumulative limit value of 500 Bq/kg was given in the Ordinance of Contaminants and Constituents in Food [3]. For group II (^{210}Pb, ^{210}Po, ^{226}Ra, ^{228}Ra, ^{230}Th, ^{232}Th and ^{231}Pa), a cumulative limit value of 50 Bq/kg was defined.

A total of 50 spice samples from Spain, South Africa, Asian countries, Turkey and India were collected on the Swiss market and analyzed with gamma-ray spectrometry. After reaching equilibrium between ^{226}Ra and ^{222}Rn (20 days), ^{226}Ra was determined via its daughter nuclides ^{214}Bi and ^{214}Pb. ^{224}Ra is in equilibrium with its daughters ^{212}Pb and ^{212}Bi. ^{228}Ra was analyzed via its daughter nuclide ^{228}Ac. The thorium nuclides ^{228}Th, ^{230}Th and ^{232}Th were analyzed with alpha spectrometry using PIPS detectors. The sources were prepared by deposition of the thorium species from acid microwave extracts onto steel disks with electrodeposition.

Pepper samples contained increased amounts of the radium and thorium nuclides. The cumulated activity of the radionuclides from group I, especially ^{224}Ra,

Radionuclide	Mean ± s.d.	Samples*	Min	Max
Pepper				
^{224}Ra	8.2 ± 6.0	5/8	2.0	16
^{226}Ra	3.4 ± 0.89	5/8	2.0	4.0
^{228}Ra	9.7 ± 7.1	6/8	3.3	23
^{228}Th	0.9 ± 0.37	3/3	0.51	1.25
^{230}Th	3.0 ± 0.72	1/3	3.0	3.0
^{232}Th	3.0 ± 3.1	2/3	0.77	5.3
Paprika				
^{224}Ra	2.0 ± 0.5	1/12	2.0	2.0
^{226}Ra	2.1 ± 0.61	3/12	1.6	2.8
^{228}Ra	<2			
^{228}Th	1.1 ± 0.85	5/5	0.27	2.4
^{230}Th	0.3 ± 0.08	1/5	0.33	0.33
^{232}Th	1.1 ± 0.79	5/5	0.37	2.5
Curries				
^{224}Ra	<2			
^{226}Ra	3.1 ± 1.7	3/11	2.0	5.0
^{228}Ra	3.3 ± 1.7	1/11	3.3	3.3
^{228}Th	1.3 ± 0.8	6/11	0.32	2.5
^{230}Th	0.4 ± 0.17	3/11	0.21	0.52
^{232}Th	1.0 ± 0.52	6/11	0.23	1.8

s.d.: Standard deviation.
**Number of positive samples/total of samples*
All data from [53, 80].

Table 9.
Investigation of spices on the Swiss market.

reached 10–32% of the limit value (500 Bq/kg). The cumulated activity of the radionuclides of group II, especially ^{226}Ra and ^{228}Ra, reached 10–26% of the limit value (50 Bq/kg) (**Table 9**). The consummation of 100 g of pepper in a year leads to a dose of about 1–2 μSv [54, 81].

6.4 Analysis of seafood and fish

Natural radionuclides cause the main radio contamination of fish and seafood. Mussels and molluscs may enrich ^{210}Po in the intestinal tract, whereas the mother nuclide ^{210}Pb is not enriched [82]. Activity concentrations of ^{210}Po range from 20 to 100 Bq/kg. In fish, the ^{210}Po level is much lower (1–20 Bq/kg) [83].

In 1998, we investigated the contamination of ^{210}Po in sea fish and mussels. The ^{210}Po was extracted with acid and microwaves at temperatures below 200°C. Then, the analytes (^{210}Po and internal standard ^{209}Po) were adsorbed onto silver disks by autodeposition in an alkaline milieu. The disks were counted with alpha PIPS detectors for 24 hours (**Table 10**).

We had to declare objections for 12 mussel and 2 fish samples (sardines). Sardines showed elevated activity of ^{210}Po. They are consumed as the whole fish, the

Radionuclide	Mean ± s.d.	Samples*	Min	Max
Fish				
^{210}Po	0.004 ± 0.005	21/	0.002	0.02
^{226}Ra	3.2 ± 6.3	54/140	0.20	43
^{228}Ra	1.3 ± 0.87	22/100	0.40	4.5
Sardines, anchovies				
^{210}Po	0.02 ± 0.03	18/19	0.001	0.11
^{226}Ra	1.7 ± 1.4	5/19	0.62	4.1
^{228}Ra	0.71 ± 0.44	6/19	0.30	1.3
Seafood				
^{210}Po	0.005 ± 0.013	11/27	0.001	0.05
^{226}Ra	2.0 ± 1.7	12/27	0.30	5.4
^{228}Ra	1.2 ± 0.10	3/27	1.1	1.2

s.d.: Standard deviation.
**Number of positive samples/total of samples.*
All data from [83, 84].

Table 10.
Investigation of fish and seafood.

intestinal tract included, similar to the consummation of mussels. This explains the higher contamination level in sardines and anchovies. In 2010, the state laboratory of Basel City undertook a second investigation with similar results. Because since 1990 the limit value for ^{210}Po in fish and seafood was raised from 10 to 150 Bq/kg (indeed, the rate of fish and seafood consummation in Switzerland is of minor relevance), no more objections had to be raised [85, 86].

7. Conclusions

In the radiation laboratory, where food and environmental samples are investigated, alpha spectrometry is a mandatory part of the instrumentation. Well-established analytical procedures exist for the analysis of polonium, radium, uranium, thorium and transuranium nuclides. Suitable radioactive sources are prepared either by co-precipitation, selective extraction or adsorption onto active surfaces. The alpha spectrometric equipment at choice are PIPS detectors, gas proportional detectors and liquid scintillation counters. Drinking water is the most important food. Periodical survey for radioactive contaminants is important to guarantee a secure consummation. Because drinking water sources may underlie seasonal variations of their activity concentrations, they must be reexamined from time to time. The dose-relevant radionuclides in our investigation of mineral waters and tap water were ^{210}Pb, ^{226}Ra and ^{228}Ra.

Additionally, there are some food categories, which may show elevated activities of alpha nuclides. Mussels and fish may enrich ^{210}Po in the gastrointestinal tract. When consuming whole fish, such as sardines or anchovies, higher amounts of ^{210}Po are taken up. This can lead to relevant doses, especially in countries where consummation of fish and seafood is a main part of the nutrition. Healing earths may contain higher amounts of ^{226}Ra and ^{228}Ra. In spices, radium species are dominant, while pepper can contain higher amounts of thorium nuclides.

Author details

Markus Zehringer
State-Laboratory Basel-City, Basel, Switzerland

*Address all correspondence to: markus.zehringer@bs.ch

References

[1] LNHB, Laboratoire National Henri Becquerel. Available from: http://www.nucleide.org/DDEP_WG/DDEPdata.htm [Accessed: November 07, 2019]

[2] Schwankner R et al. On our knowledge of radon in living spaces. Arcus. 1986;**6**:284-288

[3] The Federal Department of Home Affairs. Ordinance on Contaminants and Constituents in Food; 1995. Status: October 1, 2015

[4] Federal Department of Home Affairs. Ordinance on maximum values for contaminants (VHK); 2017. Status: May 1, 2017

[5] Federal Food Safety and Veterinary Office. Ordinance on the importation and the placing on the market of food, which is contaminated following the accident of the nuclear power station of Chernobyl; 2016. Status: May 1, 2017

[6] Federal Food Safety and Veterinary Office. Ordinance on the importation of food with origin of or consigned from Japan; 2016. Status: December 1, 2017

[7] Department of Inner Affairs. Ordinance on drinking water and water for public baths and shower facilities (TBDV); 2016. Status: May 1, 2018

[8] Council Directive 2013/51/EURATOM of October 2nd, 2013 laying down requirements for the protection of the health of the general public with regard to radioactive substances in water intended for human consumption

[9] World Health Organization. Guidelines for drinking-water quality. 4th ed. ISBN 978 92 4 154815 1. Available from: http://www.iasaude.pt/attachments/article/660/WHO_Guidelines%20for%20drinking-water%20quality.pdf [Accessed: May 25, 2019]

[10] De Goeij J, Bonardi M. How do we define the concepts specific activity, radioactive concentration, carrier, carrier-free and no-carrier-added? Journal of Radioanalytical and Nuclear Chemistry. 2005;**263**:13-18

[11] German Standard Methods for the examination of water, wastewater and sludge; general information (group A). Water quality, sampling–Part 3: Guidance on the preservation and handling of samples. Berlin: Beuth GmbH (ISO 5667-3:1994)

[12] Batley G. Collection, preparation and storage of samples for speciation analyses. In: Batley G, editor. Trace Elements Speciation: Analytical Methods and Problems. Boca Raton, Florida: CRC Press, Inc; 1990. ISBN: 0-8493-4712-2

[13] Batley G, Gardner D. Sampling and storage of natural waters for trace metal analysis. Water Research. 1977;**11**: 745-756

[14] Seiler H. Some recommendations for the specimen collection of biological materials for analysis. In: Seiler H, Sigel H, Sigel A, editors. Handbook on Toxicity of Inorganic Compounds. New York. Basel: Marcel Dekker; 1988. p. 41. ISBN: 0-8247-7727-1

[15] Robertson D. Role of contamination in trace element analysis of sea water. Analytical Chemistry. 1968;**40**: 1067-1072

[16] Versieck J, Vanballenberghe L. Collection, transport and storage of biological samples for the determination of trace metals. In: Seiler H, Sigel H, Sigel A, editors. Handbook on Metals in Clinical and Analytical Chemistry. Basel: Marcel Dekker; 1994. pp. 39-42. ISBN: 0-8247-9094-4

[17] German Standard Methods for the examination of water, waste water and sludge; physical and physical-chemical parameters (group C). Determination of radium-226-concentration in potable water, ground water, surface water and wastewater. Berlin: Beuth GmbH; 1994

[18] McDowell J, McDowell B. Liquid Scintillation Alpha Spectrometry. Boca Raton, Florida: CRC Press; 1994. ISBN: 0-8493-5288-6

[19] Aupiais J, Fayolle C, Gilbert P, Dacheux N. Determination of ^{226}Ra in mineral drinking waters by a liquid scintillation with rejection of β-γ emitters. Analytical Chemistry. 1998;**70**: 2353-2359

[20] Aupiais J. Radium measurement in water samples by α-liquid scintillation counting with α/β discrimination. Analytica Chimica Acta. 2005;**532**: 199-207

[21] Burnett W, Tai W. Determination of radium in natural waters by a liquid scintillation. Analytical Chemistry. 1992;**64**:1691-1697

[22] Escobar V, Tome F, Lozano J, Sanchez A. Extractive procedure for uranium determination in water samples by liquid scintillation counting. Applied Radiation and Isotopes. 1998; **49**:875-883

[23] Aupiais J. Rapid determination of uranium activity and concentration in water by alpha liquid scintillation with α/β discrimination. Analytica Chimica Acta. 2004;**517**:221-228

[24] Leyba J, Vollmar H, Fjeld R, Devol T, Brown D, Cadieux J. Evaluation of a direct extraction/liquid scintillation counting technique for the measurement of uranium in water. Journal of Radioanalytical and Nuclear Chemistry. 1995;**194**:337-344

[25] Véronneau C, Aupiais J, Dacheux N. Selective determination of polonium by photon electron rejecting alpha liquid scintillation (PERALS® system). Analytica Chimica Acta. 2000;**415**: 229-238

[26] Ordela. Available from: http://www.ordela.com [Accessed: May 21, 2019]

[27] Möbius S, Kamolchote K, Rakotoarisoa T. Extractive methods for fast radium analysis. In: Möbius S, Noakes J, Schönhofer F, editors. International Conference on Advances in Liquid Scintillation Spectrometry Tucson, Arizona: Radiocarbon. 2001. pp. 283-292

[28] Dacheux N, Aupiais J, Courson O, Aubert C. Comparison and improvement of the determination of actinides with low activities using several alpha liquid scintillation spectrometers. Analytical Chemistry. 2000;**72**:3150-3157

[29] Landstetter C, Hiegesberger B, Sinojmeri M, Katzlberger C. Determination of ^{210}Pb and ^{210}Po in water using the extractive scintillation cocktail POLEX™. Applied Radiation and Isotopes. 2014;**913**:76-81

[30] Aupiais J, Dacheux N, Thomas A, Matton S. Study of neptunium measurement by alpha liquid scintillation with rejection of β-, γ-emitters. Analytica Chimica Acta. 1999;**398**:205-2018

[31] Hamanaka S, Shizuma K, Wen X, Iwatani K, Hasai H. Radon concentration measurement in water by means of α liquid-scintillation spectrometry with a PERALS spectrometer. Nuclear Instruments & Methods. 1998;**410**:313-318

[32] Hashimoto T, Sato K, Yoneyama Y, Fukuyama N. Simultaneous determination of environmental

α-radionuclides using liquid scintillation counting combined with time interval analyses (TIA) and pulse shape discrimination (PSD). Journal of Radioanalytical and Nuclear Chemistry. 1997;**222**:109-116

[33] Zehringer M, Abraham J, Kammerer F, Syla V, Wagmann M. A rapid extraction procedure for the determination of strontium-90 in water samples. Journal of Chemistry and Chemical Engineering. 2017;**11**: 116-123

[34] Wallner G. Thorium determination by liquid scintillation counting using an extractive cocktail. Environment International. 1996;**22**:S101-S103

[35] Ayranov M, Wacker L, Krähenbühl U. Determination of uranium and thorium in complex matrices by two solvent extraction separation techniques and photon electron rejecting alpha liquid spectrometry. Radiochimica Acta. 2001; **89**:823-827

[36] Leeuwen F, Verboom W, Reinhoudt D. Selective extraction of naturally occurring radioactive Ra^{2+}. Chemical Society Reviews. 2005;**34**: 753-761

[37] Jia G, Jia J. Determination of radium isotopes in environmental samples by gamma spectrometry, liquid scintillation counting and alpha spectrometry: A review of analytical methodology. Journal of Environmental Radioactivity. 2012;**106**:98-119

[38] Xarchoulakos D, Kehagia K. A study of various self-deposition solutions for ^{210}Po analysis in tap water, Journal of Radioanalytical and Nuclear Chemistry 2019;**319**:419-424

[39] Carvalho F, Fowler S. An experimental study on the bioaccumulation and turnover of polonium-210 and lead-210 in marine shrimp. Marine Ecology Progress Series. 1993;**102**:125-133

[40] Yamamoto M et al. Polonium-210 and lead-210 in marine organisms: Intake levels for Japanese. Journal of Radioanalytical and Nuclear Chemistry. 1994;**178**:81-90

[41] Clayton R, Bradley E. A cost-effective method for the determination of ^{210}Po and ^{210}Pb in environmental materials. Science of the Total Environment. 1995;**173**:22-28

[42] Moore W, Reid D. Extraction of radium from natural waters using manganese-impregnated acrylic fibers. Journal of Geophysical Research. 1973; **78**:8880-8886

[43] Glöbel B, Berlich J. Eine einfache und schnelle Methode zur Bestimmung von ^{226}Ra in wässrigen Proben. In: Proc. Fachgespräch Überwachung der Umweltradioaktivität; Karlsruhe. 1983

[44] Surbeck H. Determination of natural radionuclides in drinking water: A tentative protocol. Science of the Total Environment. 1995;**173–174**:91-99

[45] Eikenberg J, Tricca A, Vezzu G, Bajo S, Ruethi M, Surbeck H. Determination of ^{228}Ra, ^{226}Ra and ^{224}Ra in natural water via adsorption on MnO_2-coated discs. Journal of Environmental Radioactivity. 2001;**54**:109-131

[46] Surbeck H. Alpha spectrometry sample preparation using selective adsorbing thin films. Applied Radiation and Isotopes. 2000;**53**:97-100

[47] Triskem. Available from: http://www.triskem-international.com/scripts/files/59d1f4fc2acb49.99908156/tki_4_en.pdf [Accessed: 2019-05-21]

[48] Chiariza R, Horwitz E, Alexandratos S, Gula M. Diphonix resin: A review of its properties and

applications. Separation Science and Technology. 1997;**32**:1-35

[49] Surbeck H. Personal Communication. (e-mail from January 31st 2019)

[50] Mitchell R. Electrodeposition of actinide elements at tracer concentrations. Analytical Chemistry. 1960;**32**:326-328

[51] Talvitie N. Electrodeposition of actinides for alpha spectrometric determination. Analytical Chemistry. 1972;**44**:280-283

[52] Percival D, Martin D. Sequential determination of radium-226, radium-228, actinium-227 and thorium isotopes in environmental and process waste samples. Analytical Chemistry. 1974;**46**:1742-1749

[53] Harduin J, Peleau B, Levavasseur D. Analytical determination of actinides in biological samples. Radioprotection. 1996;**31**:229-245

[54] Thimmaiah D, Zehringer M. Determination of actinides in spices by means of alpha spectrometry after enrichment by an electrodepositon procedure. Chimia. 2012;**66**:493

[55] Bajo S. Calibration of an Alpha-Spectrometer with Electrodeposited Uranium Sources. Paul Scherrer Institut Technische Mitteilung; Würenlingen; 1996; TM-23-96-14

[56] Lally A, Glover K. Source preparation in alpha spectrometry. Nuclear Instruments and Methods in Physics Research. 1984;**223**:259-265

[57] Parker W, Bildstein H, Getoff N. Molecular plating I. A rapid and quantitative method for the electrodepositon of thorium and uranium. Nuclear Instruments & Methods. 1964;**26**:55-60

[58] Crespo M. A review of electrodeposition methods for the preparation of alpha-radiation sources. Applied Radiation and Isotopes. 2012; **70**:210-215

[59] Frindik O, Heilgeist M, Kalus W, Schelenz R. Verfahren zur alphaspektrometrischen bestimmung von plutonium, uran, americium, curium und thorium in Lebensmitteln. In: Messanleitungen Umweltradioaktivität. Munich: Urban& Fischer; 2000

[60] Hindman F. Neodymium fluoride mounting for alpha spectrometric determination of uranium, plutonium and americium. Analytical Chemistry. 1983;**55**:2460-2461

[61] Sill C, Williams R. Preparation of actinides for alpha spectrometry without electrodeposition. Analytical Chemistry. 1981;**53**:421-415

[62] Sawant P et al. Alpha source preparation of actinides by micro-precipitation. Journal of Radioanalytical and Nuclear Chemistry. 2018;**319**: 109-113

[63] 3M Science. Applied to life. Available from: https://www.3m.com, https://www.sopachem.com/analytical/wp-content/uploads/2014/07/3M-Empore-Radium-RAD-Disks.pdf [Accessed: May 21, 2019]

[64] Möbius S, Kamolchote K, Rakotoarisoa T. Extractive methods for fast radium analysis. In: Möbius S, Noakes J, Schönhofer F, editors. International Conference on Advances in Liquid Scintillation Spectrometry LSC 2001. Radiocarbon: Tucson. pp. 253-259. ISBN: 0-9638314-4-5

[65] L'Annunziata M, editor. Handbook of Radioactivity Analysis. 2nd ed. San Diego: Academic Press; 2003. ISBN: 0-12-436603-1

[66] Knoll G. Radiation Detection and Measurement. 3rd ed. New York: John

Wiley & Sons; 1999. ISBN: 0-471-81504-7

[67] Ortec. Available from: http://www.ortec-online.com/ [Accessed: May 21, 2019]

[68] Mirion (former Canberra). Available from: http://www.canberra.com/products/detectors/ [Accessed: May 21, 2019]

[69] Möbius S, Noakes J, Schönhofer F. International Conference on Advances in Liquid Scintillation Spectrometry LSC. Tucson: Radiocarbon. 2001. 449p. ISBN: 0-9638314-4-5

[70] Frenzel E, Kossert K, Oikari T, Otto R, Wisser S. Die Messung von ^{89}Sr/^{90}Sr und ^{90}Sr/^{90}Y mittels TDCR-Cerenkov-Zählung. Strahlenschutzpraxis. 2013;**1**:26-33

[71] Zehringer M. Gamma-ray spectrometry and the investigation of environmental and food samples. In: Maghraby A, editor. New Insights on Gamma Rays. Croatia: InTech; 2017. pp. 3-27. ISBN 978-953-51-3162-5

[72] The Federal Department of Home Affairs. Ordinance on Beverages; 2016. Status: May 1, 2017

[73] Homepage of the Swiss Association of Mineral Water Sources and Softdrink Producers. Available from: https://mineralwasser.swiss [Accessed: May 21, 2019]

[74] Bundesministerium für Umwelt, Naturschutz, Bau und Reaktorsicherheit (BMUB). 2017. Leitfaden zur Untersuchung und Bewertung von radioaktiven Stoffen im Trinkwasser bei der Umsetzung der Trinkwasserverordnung. Empfehlungen von BMUB, BMG, BfS, UBA und den zuständigen Landesbehörden sowie DVGW und BDEW. Available from: https://doris.bfs.de/jspui/bitstream/urn:nbn:de:0221-2017020114224/5/20180530_Leitfaden%20Trinkwasser_mit_Formblaettern.pdf [Accessed: May 25, 2019]

[75] Aellen T, Ch W, Görlich W, Umbricht O. Natürliche Radionuklide der Uran- und Thorium Zerfallsreihe in Mineralwässern. In: Federal Office of Public Health, editor. Environmental Radioactivity and Radiation Doses in Switzerland. 1990. Bern: Office of Public Health, B.3.5.10.-16. ISBN: 3-905235-04-8; Available from: https://www.bag.admin.ch/bag/de/home/das-bag/publikationen/taetigkeitsberichte/jahresberichte-umweltradioaktiviaet.html [Accessed: May 25, 2019]

[76] Zehringer M. Uranium, radium, thorium and tritium in mineral and table waters. In: Annual Report of the State-Laboratory Basel-City. 2006. pp. 83-86

[77] Zehringer M. Organic micropollutants, heavy metals and radioactivity in mineral waters. In: Annual Report of the State-Laboratory of Basel-City. 2007. pp. 70-72. Available from: https://www.kantonslabor.bs.ch/berichte/jahresberichte.html [Accessed May 25, 2019]

[78] Swiss Federal Council. Radiological protection ordinance (RPO); April 26, 2017. Status: February 1, 2019

[79] Zehringer M. Radiological investigation of siliceous earths. In: Annual Report of the State-Laboratory of Basel-City. 2008. pp. 79-82. Available from: https://www.kantonslabor.bs.ch/berichte/jahresberichte.html [Accessed May 25, 2019]

[80] Zehringer M. Radionuclides in healing earths. In: Annual Report of the State-Laboratory of Basel-City. 2010. pp. 70-73. Available from: https://www.kantonslabor.bs.ch/berichte/jahresberichte.html [Accessed May 25, 2019]

[81] Zehringer M. Sudan dyes, heavy metal traces and radio contamination of

spicery. In: Annual Report of the State-Laboratory of Basel-City. 2006. pp. 89-91. Available from: https://www.kantonslabor.bs.ch/berichte/jahresberichte.html [Accessed May 25, 2019]

[82] Cherry R, Shannon L. The alpha radioactivity of marine organisms. Atomic Energy Review. 1974;**12**:3-45

[83] Froidevaux P, Dell D, Tossell P. Radionuclides in foodstuffs and food raw material. In: Pöschl M, Nollet L, editors. Radionuclide Concentrations in Food and the Environment. New York: CRC Taylor & Francis; 2007. pp. 239-243. ISBN: 0-8493-3594-9. Ch. 8

[84] Zehringer M. Radioactivity in food: Experiences of the food control authority of Basel-city since the Chernobyl accident. In: Monteiro W, editor. Radiation Effects in Materials. Croatia: InTech; 2016: 132-160. Available from: http://dx.doi.org/10.5772/62460

[85] Zehringer M. Gamma nuclides, polonium-210 and heavy metal traces in fish and seafood. In: Annual Report of the State-Laboratory of Basel-City. 2010. pp. 57-59. Available from: https://www.kantonslabor.bs.ch/berichte/jahresberichte.html [Accessed May 25, 2019]

[86] Zehringer M. Radioactivity in sardines, anchovies and tuna. In: Annual report of the State-Laboratory of Basel-City. 2011. pp. 31-33. Available from: https://www.kantonslabor.bs.ch/berichte/jahresberichte.html [Accessed May 25, 2019]

Chapter 9

Analysis of Radioactive Elements in Testes of Large Japanese Field Mice Using an Electron Probe Micro-Analyser after the Fukushima Accident

Takuya Ohdaira, Kanna Meguro, Kazuki Komatsu, Rina Syoji, Yohei Fujishima, Valerie Swee Ting Goh, Kosuke Kasai, Kentaro Ariyoshi, Akifumi Nakata, Yusuke Urushihara, Kazuma Koarai, Yasushi Kino, Tsutomu Sekine, Masatoshi Suzuki, Atsushi Takahasi, Yoshinaka Shimizu, Hisashi Shinoda, Mitsuaki A. Yoshida, Manabu Fukumoto, Hideaki Yamashiro and Tomisato Miura

Abstract

The Fukushima Daiichi nuclear power plant (FDNPP) accident drew global attention to the health risks of radiation exposure. The large Japanese field mice (*Apodemus speciosus*) are rodents endemic to, and distributed throughout, Japan. This wild rodent live in and around the ex-evacuation zone on the ground surface and/or underground. In this study, we evaluated the effect of chronic radiation exposure associated with FDNPP accident on the testes of large Japanese field mice. Morphological analysis and electron-prove X-ray microanalysis (EPMA) was undertaken on the testes. Morphological analysis of testes based on H&E staining showed that the spermatogenesis was observed normally in the breeding season of wild mice in the heavily contaminated area. However, caesium (Cs) was not detected in all testes of wild mice from FDNPP ex-evacuation zone. In conclusion, even if the testes and the process of spermatogenesis are hypersensitive to radiation, we could not detect radiation effects on the spermatogenesis and Cs in the examined large Japanese field mice testes following chronic radiation exposure associated with the FDNPP accident.

Keywords: EPMA analysis, Fukushima nuclear power plant accident, testis, wild mice

1. Introduction

The Fukushima Daiichi nuclear power plant (FDNPP) accident drew global attention to the health risks of radiation exposure. We have established an archive system composed of livestock and wild animals within a 20 km radius from FDNPP, that is, the ex-evacuation zone of the FDNPP accident [1–13]. This system provides critical information for the understanding of environmental pollution, biodistribution, radionuclide metabolism, dose evaluation, and the biological effects of internal and external exposure to radiation caused by nuclear disasters. In particular, experimental studies of low-dose rate (LDR) radiation exposure induced effects on spermatogenesis, along with indications from the nuclear disaster in Fukushima, will provide a more comprehensive radiobiological understanding of response mechanisms leading to improved accuracy in the estimation of human reproduction and health risk [14].

The large Japanese field mice (*Apodemus speciosus*) are rodents endemic to, and distributed throughout, Japan [15]. This wild rodent is appropriate for use as a reference animal of the ecosystem. Large Japanese field mice live in and around the ex-evacuation zone on the ground surface and/or underground. Hence, they are exposed to high levels of external radiation. Furthermore, they eat contaminated food and drink contaminated water. Consequently, they are directly affected by radioactive substances. Therefore, these mice can serve as a model to study the effect of radiation exposure, while also serving as a reference animal for the surrounding ecosystem.

Electron probe X-ray microanalysis (EPMA) is a powerful tool used to detect trace amounts of chemical elements in single cells and tissues [16]. This method measures the characteristic X-ray spectra of specific elements in samples using an accelerated electron beam. We previously investigated the effect of chronic LDR exposure to ^{134}Cs and ^{137}Cs on the testis of euthanised bulls, boars, and inobutas from the evacuation zone [3, 7].

Discharge of ^{134}Cs and ^{137}Cs that emit γ- and β-rays is of primary concern, because they were released in a large amount and have a long half-life. In this study, we evaluated the heavy contamination levels of LDR effects of ^{134}Cs and ^{137}Cs (between 4848 and 70,200 Bq/kg) on the large Japanese field mice after the FDNPP accident.

2. Materials and methods

2.1 Collections of large Japanese field mice

The study protocol followed laboratory animal care guidelines, and all procedures were conducted in accordance with the guideline of the Ethics Committee for Care and Use of Laboratory Animals for Research of Niigata University, Japan (approval number: H2611). Large Japanese field mice were captured using Sherman traps (H.B. Sherman Traps, Inc., Tallahassee, FL, USA) at three sites, Akogi, Ide, and Omaru of Namie town in the ex-evacuation zone of the FDNPP accident in November 2012, April 2013, and April 2016 (**Figure 1**). Control large Japanese mice were captured using Sherman traps in May 2012, November 2015, and April 2016 in Aomori Prefecture, and April and May 2016 in Niigata Prefecture. The ambient dose rate was measured at the sampling sites using NaI (Tl) scintillation survey meter TCS-171B (Hitachi Aloka Medical, Ltd., Tokyo, Japan) at the height of 1 m. The measurements were expressed as micrograys per hour at 1 m above the ground.

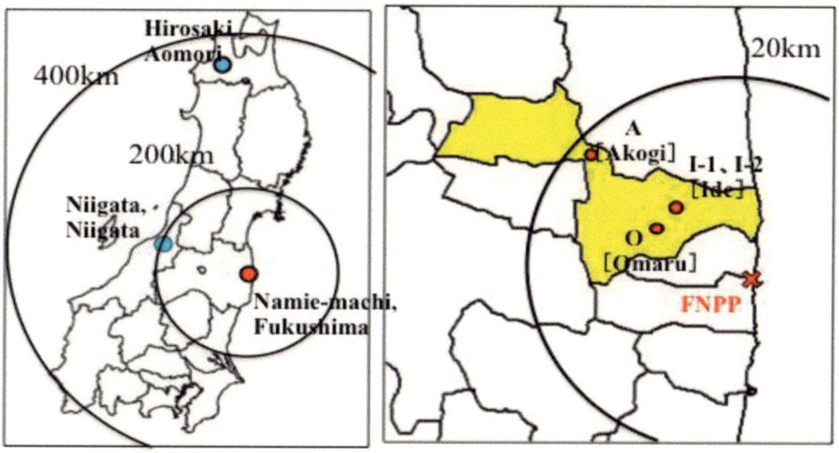

Figure 1.
Sampling site of in Namie town, Niigata and Aomori. Akogi, Ide and Omaru of Namie town in the ex-evacuation zone of the FDNPP accident is shown in yellow.

2.2 Measurement of radioactivity

Radioactivities of the organ samples were determined via gamma-ray spectrometry using high-purity germanium (HPGe) detector (GEM40P4-83, Ortec Co., Oak Ridge, TN, USA) as described previously [10]. The duration of the measurement varied from 110,600 to 663,400 s, depending on the radioactivity of the sample. Absolute efficiency of the detector was determined with the standard point sources of ^{137}Cs (10 kBq, CS402) and ^{152}Eu (10 kBq, EU402, Japan Isotope Association, Tokyo, Japan). The samples were placed in a small space (1 mm thick and 6 mm diameter) which is the same size as the standard point sources. A nuclide was identified when its characteristic photopeak 3σ above the baseline observed in the spectrum. The activities due to decay were corrected to the sampling dates.

2.3 Morphological assessment of testes cells

The testes were fixed in Bouin's solution, embedded in paraffin, and stained using haematoxylin and eosin (H&E), according to standard protocols, as described by Takino et al. [11]. Subsequently, the testes were briefly dehydrated in different concentrations of alcohol. The testes were made transparent by using toluene, and then, then they were embedded in paraffin and cut into 5 μm-thick sections before staining.

2.4 Electron probe X-ray microanalysis

Qualitative analysis: An analytical method was used to investigate the composition of the sample to be analysed. From ^{6}B to ^{92}U can be measured with a combination of analytical crystal to be used. The analysis conditions were as follows: voltage was set to 15 kV, beam current was 100 nA, beam size was minimum, sample current was 92.8 nA, and time 30 ms/point (**Figure 2**).

Elements analysis: Chemical trace analyses of caesium (Cs), sulphur (S), and nitrogen (N) in the testes were performed using a Shimadzu 1720HT electron probe micro-analyser (Shimadzu Corporation, Tokyo, Japan) equipped for X-ray spectrometry and specifically adapted for the examination of ultrathin sections. Accordingly, 3 μm of each testis section was placed on the carbon plate, and

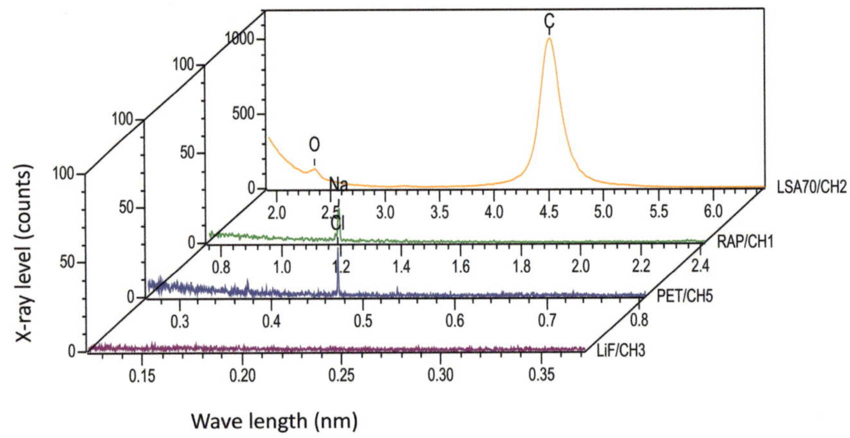

Figure 2.
Result of qualitative analysis by EPMA.

subsequently, each section was carbon coated for the electrification of samples (Biopathology Institute Co., Ltd., Oita, Japan). For the analysis, the voltage of the electron microscope was set to 15 kV, and the electron beam rate was set to 100 nA. Other parameters were beam size minimum × region (260 × 195 μm) and time (30 ms/point). The sections were viewed as secondary electron images, and chemical elemental mapping was performed. We performed EPMA analysis duplicate including test analysis.

3. Results and discussion

To date, low-dose radiation effects on physiological processes including spermatogenesis remain unclear. Further studies are required to confirm these low-dose radiation effects [14].

In the present study, we examined the effects of LDR exposure associated with the FDNPP accident on the testes of large Japanese field mice from different contaminated areas in the ex-evacuation zone, at Namie town in Fukushima. The ambient dose rate at Akogi was 26.9 μGy/h in November 2012, and 15.2 μGy/h in April 2013. The dose rate at Omaru was 12.3 μGy/h in April 2016. The dose rate at Ide was 16.4 μGy/h in April 2013, and 5.3 μGy/h in April 2016 (**Table 1**).

The ^{134}Cs and ^{137}Cs radioactivity concentrations (Bq/kg) in large Japanese field mice organ samples were detected via gamma-ray spectrometry by using an HPGe detector (**Table 1**). The total radioactivity concentrations of ^{134}Cs and ^{137}Cs in large Japanese field mice organ samples from Omaru were 2510, 2750, 3860, and 37,630 Bq/kg. Those from Ide were 10,820 and 16,550 Bq/kg, and this level is highly contaminated for the large Japanese field mice in the ex-evacuation zone.

Okano et al. [17] reported that, although the concentrations of ^{134}Cs and ^{137}Cs in wild mice from Fukushima exceeded 4000 Bq/kg, there were no significant differences in the frequencies of apoptotic cells or morphologically abnormal sperm when compared with wild mice from the non-contaminated control area. However, Kawagoshi et al. [18] reported that the average frequencies of chromosomal aberrations in splenic lymphocytes of animals living in the heavily contaminated (approximately 3 mGy/day) area of Fukushima were higher than those of animals from the non-contaminated, slightly contaminated (approximately 0.03 mGy/day), and moderately contaminated (approximately 1 mGy/day) areas. Moreover, the

Area	Large Japanese field mice			Sampling date	Ambient dose rate (μGy/h)	Body weight (g)	Radioactive concentration of ^{134}Cs and ^{137}Cs		
	No.	ID	Site				^{134}Cs (Bq/kg)	^{137}Cs (Bq/kg)	Total (Bq/kg)
Fukushima	1	215	Akogi	11/6/2012	26.9	23.5	–	–	–
	2	260		19/04/2013	15.2	32.4	–	–	–
	3	572	Omaru	12/4/2016	12.3	36.2	580	2880	3460
	4	575		12/4/2016	12.3	29.2	500	2250	2750
	5	590		12/4/2016	12.3	10.8	690	3170	3860
	6	594		12/4/2016	12.3	28.1	460	2050	2510
	7	595		12/4/2016	12.3	44.5	6360	31,270	37,630
	8	257	Ide	19/04/2013	16.4	30.2	–	–	–
	9	596		12/4/2016	5.3	50.1	1960	8860	10,820
	10	597		12/4/2016	5.3	43.5	2990	13,560	16,550
Aomori (control)	11	150	Hirosaki	29/05/2012	–	37	–	–	–
Niigata (control)	12	2721	Kakuta	20/11/2015	–	34.2	–	–	–
	13	2811		18/04/2016		43.8			

Table 1.
Individual information for large Japanese field mice.

aberration frequency in individual wild mice tended to increase with the estimated dose rates and accumulated doses. Takino et al. [11] reported that enhanced spermatogenesis occurred in large Japanese field mice living in and around the ex-evacuation zone of FDNPP. It remains to be elucidated whether the phenomenon, which is attributable to chronic LDR exposure, has a beneficial or adverse effect on large Japanese field mice.

Morphological analysis of testes based on H&E staining showed that the stages of the seasonal reproductive cycle were classified into reproductive, non-reproductive, and transition periods (**Figure 3A–D**; 1). During the reproductive seasons of the large Japanese field mice from Ide, spermatogonia, primary spermatocyte, secondary spermatocyte, and sperm were observed (**Figure 3B**). Interestingly, spermatogenesis was also observed normally in the breeding season of wild mice in the heavily contaminated area of Omaru (**Figure 3C**). Moreover, it was confirmed that the regression of sperm and seminiferous tubules during the non-breeding season of the wild mice in the most heavily contaminated area of Akogi were normally observed (**Figure 3D**).

Figure 3A–D (images 4–6) presents the phase maps obtained using the EPMA indicating micro-constituent concentrations namely, Cs, S and N. Colour imaging rapidly and effectively facilitates the overall analysis of the composite structure; specifically, decreasing levels of metal distribution are indicated from red to blue. Cs was not detected in all testes of wild mice from Ide, Akogi, and Omaru (**Figure 3A–D**: images 4). In the breeding samples, sulphur was detected inside seminiferous tubules, especially in sperm and was detected around the seminiferous tubules in the non-breeding seasons (**Figure 3A–D**: images 5). Nitrogen was detected inside both the seminiferous tubules and membranes (**Figure 3A–D**: images 6).

In conclusion, even if the testes and the process of spermatogenesis are hypersensitive to radiation, there were no significant radiation effects on the

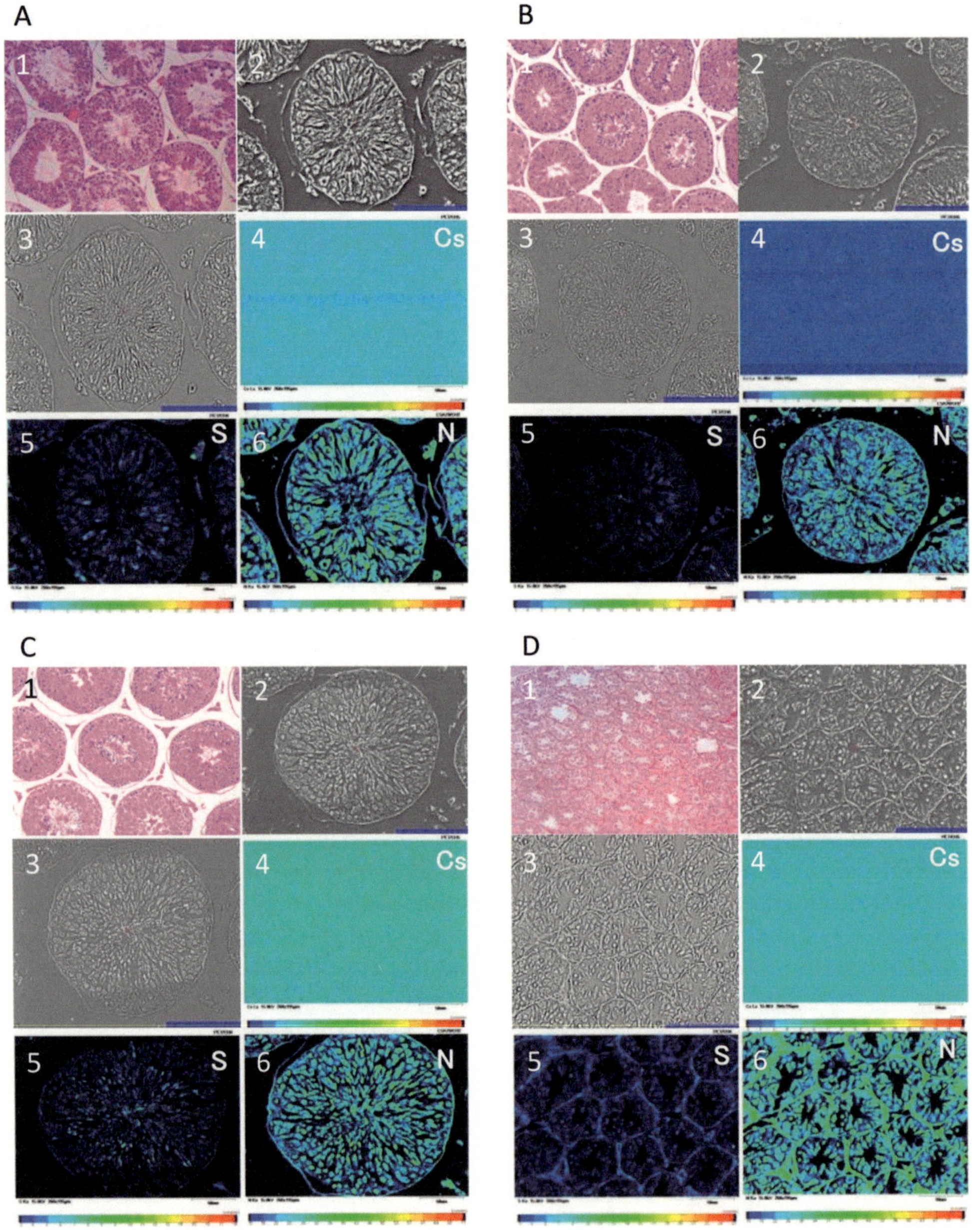

Figure 3.
Elements analysis of large Japanese field mice testis. A. Control (ID 2811), B. Ide (ID 596), C. Akogi (ID 215), and D. Omaru (ID 595). (1) H & E staining images of testis. (2) Stereo-microscopy images. (3) Composite backscattered microscopy images. (4) Colour map images of Cs (caesium). (5) Colour map images of S (sulphur). (6) Colour map images of N (nitrogen).

spermatogenesis and Cs in the examined large Japanese field mice testes following chronic LDR radiation exposure associated with the FDNPP accident.

Acknowledgements

This work was supported by the Japan Society for the Promotion of Science, and entrusted to the Japan Atomic Energy Agency (JAEA) by the Ministry of Education, Culture, Sports, Science, and Technology of Japan (MEXT).

DOI: http://dx.doi.org/10.5772/intechopen.84634

Author details

Takuya Ohdaira[1], Kanna Meguro[1], Kazuki Komatsu[1], Rina Syoji[1], Yohei Fujishima[2], Valerie Swee Ting Goh[2], Kosuke Kasai[2], Kentaro Ariyoshi[3], Akifumi Nakata[4], Yusuke Urushihara[5], Kazuma Koarai[6], Yasushi Kino[6], Tsutomu Sekine[6,7], Masatoshi Suzuki[8], Atsushi Takahasi[9], Yoshinaka Shimizu[9], Hisashi Shinoda[9], Mitsuaki A. Yoshida[3], Manabu Fukumoto[10], Hideaki Yamashiro[1]* and Tomisato Miura[2,3]*

1 Graduate School of Science and Technology, Niigata University, Niigata, Japan

2 Graduate School of Health Sciences, Hirosaki University, Aomori, Japan

3 Institute of Radiation Emergency Medicine, Hirosaki University, Aomori, Japan

4 Faculty of Pharmacy, Hokkaido University of Science, Hokkaido, Japan

5 Graduate School of Medicine, Tohoku University, Sendai, Japan

6 Graduate School of Chemistry, Tohoku University, Sendai, Japan

7 Institute for Excellence in Higher Education, Tohoku University, Sendai, Japan

8 Institute for Disaster Reconstruction and Regeneration Research, Tohoku University, Sendai, Japan

9 Graduate School of Dentistry, Tohoku University, Sendai, Japan

10 Department of Molecular Pathology, Tokyo Medical University, Tokyo, Japan

*Address all correspondence to: hyamashiro@agr.niigata-u.ac.jp and tomisato@hirosaki-u.ac.jp

References

[1] Fukuda T, Kino Y, Abe Y, Yamashiro H, Kuwahara Y, Nihei H, et al. Distribution of artificial radionuclides in the abandoned cattle in the evacuation zone of the Fukushima Daiichi nuclear power plant. PLoS One. 2013;**8**, e54312

[2] Isogai E, Kino Y, Abe Y, Yamashiro H, Shinoda H, Fukuda T, et al. Distribution of radioactive cesium in ostrich (*Struthio Camelus*) after the Fukushima Daiichi nuclear power plant accident. Radiation Emergency Medicine. 2013;**2**:68-71

[3] Yamashiro H, Abe Y, Fukuda T, Kino Y, Kawaguchi I, Kuwahara Y, et al. Effects of radioactive caesium on bull testes after the Fukushima nuclear plant accident. Scientific Reports. 2013;**3**:2850

[4] Hosoda M, Tokonami S, Tazoe H, Sorimachi A, Monzen S, Osanai M, et al. Activity concentrations of environmental samples collected in Fukushima Prefecture immediately after the Fukushima nuclear accident. Scientific Reports. 2013;**3**:2283

[5] Fukuda T, Kino Y, Abe Y, Yamashiro H, Kobayashi J, Shimizu Y, et al. Cesium radioactivity in peripheral blood is linearly correlated to that in skeletal muscle: Analyses of cattle within the evacuation zone of the Fukushima Daiichi nuclear power plant. Animal Science Journal. 2015;**86**:120-124

[6] Takahashi S, Inoue K, Urushihara Y, Hayashi G, Kino Y, Sekine T, et al. A comprehensive dose evaluation project concerning animals affected by the Fukushima Daiichi nuclear power plant accident: Its setup and progress. Journal of Radiation Research. 2015;**56**(S1):i36-i41

[7] Yamashiro H, Abe Y, Hayashi G, Urushihara Y, Kuwahara Y, Suzuki M, et al. Electron probe X-ray microanalysis of boar and inobuta testes after the Fukushima accident. Journal of Radiation Research. 2015;**56**(S1):i42-i47

[8] Fukuda T, Hiji M, Kino Y, Abe Y, Yamashiro H, Kobayashi J, et al. Software development for estimating the cesium radioactivity in skeletal muscle from that in blood of cattle. Animal Science Journal. 2016;**87**:842-847

[9] Koarai K, Kino Y, Takahashi A, Suzuki T, Shimizu Y, Chiba M, et al. ^{90}Sr in teeth of cattle abandoned in evacuation zone: Record of pollution from the Fukushima-Daiichi nuclear power plant accident. Scientific Reports. 2016;**6**:24077

[10] Urushihara Y, Kawasumi K, Endo S, Tanaka K, Hirakawa Y, Hayashi G, et al. Analysis of plasma protein concentrations and enzyme activities in cattle within the ex-evacuation zone of the Fukushima Daiichi nuclear plant accident. PLoS One. 2016;**11**:e0155069

[11] Takino S, Yamashiro H, Sugano Y, Fujishima Y, Nakata A, Kasai K, et al. Analysis of the effect of chronic and low-dose radiation exposure on spermatogenic cells of male large Japanese field mice (*Apodemus speciosus*) after the Fukushima Daiichi nuclear power plant accident. Radiation Research. 2017;**187**:161-168

[12] Koarai K, Kino Y, Takahashi A, Suzuki T, Shimizu Y, Chiba M, et al. ^{90}Sr specific activity of teeth of abandoned cattle after the Fukushima accident—Teeth as an indicator of environmental pollution. Journal of Environmental Radioactivity. 2018;**183**:1-6

[13] Ariyoshi K, Miura T, Kasai K, Akifumi N, Fujishima Y, Yoshida MA. Radiation-induced bystander effect in large Japanese field mouse (*Apodemus speciosus*) embryonic cells. Radiation and Environmental Biophysics. 2018;**57**:223-231

[14] Fukunaga H, Butterworth KT, Yokoya A, Ogawa T, Prise KM. Low-dose radiation-induced risk in spermatogenesis. International Journal of Radiation Biology. 2017;**93**:1291-1298

[15] Suzuki H, Yasuda SP, Sakaizumi M, Wakana S, Motokawa M, Tsuchiya K. Differential geographic patterns of mitochondrial DNA variation in two sympatric species of Japanese wood mice, *Apodemus speciosus* and *A. argenteus*. Genes & Genetic Systems. 2004;**79**:165-176

[16] Pogorelov AG, Budantsev AY, Pogorelova VN. Quantitative electron probe microanalysis of acetylcholinesterase activity in rat brain sections. Journal of Histochemistry and Cytochemistry. 1993;**41**:1795-1800

[17] Okano T, Ishiniwa H, Onuma M, Shindo J, Yokohata Y, Tamaoki M. Effects of environmental radiation on testes and spermatogenesis in wild large Japanese field mice (*Apodemus speciosus*) from Fukushima. Scientific Reports. 2016;**6**:23601

[18] Kawagoshi T, Shiomi N, Takahashi H, Watanabe Y, Fuma S, Doi K, et al. Chromosomal aberrations in large Japanese field mice (*Apodemus speciosus*) captured near Fukushima Dai-ichi nuclear power plant. Environmental Science & Technology. 2017;**51**:4632-4641

Chapter 10

Environmental Radiation: Natural Radioactivity Monitoring

Isam Salih Mohamed Musa

Abstract

People are continuously exposed to ionizing radiation from many sources, including natural radioactive substances that are produced in the atmosphere and on Earth, in addition to radionuclides manufactured for various applications. Exposures vary among different places depending on many parameters. There are regions with considerably high natural background radiation while most places classified as low to medium levels. This noticeable interest pronounced worldwide is radioactivity monitoring and wide surveys by many countries. This is useful for the assessment of public as well as creating baseline if changes in the levels due to activities and practices take place. Good management of protecting the environment is to establish baseline data of high quality with efficient measurements. Natural radioactivity monitoring in view of radiation and environmental protection is presented in this chapter. It will shed the light on rapid methods describing the measurement of radionuclides and assessment of exposure to human. It presents a summary of methods of radiation dose calculations that individual may be exposed with some numerical examples. The chapter also presents methods for predicting the spatial distribution of radiological quantities using geographical information system. Environmental measurements may be costly and time-consuming practices; hence, thoughts to reduce time force itself in this chapter.

Keywords: environmental radioactivity, monitoring, gamma spectrometry, ambient dose, absorbed dose, GIS, geostatistics

1. Introduction

Our planet, the earth, is a wonderful place and has been suitable to live on its surface for thousands of years; it obliges us to preserve and nurture it. As investment volumes continue to grow in the globalized economy, environmental shadows are intersecting more and more on this planet. The concept of sustainable development, which generally means meeting the needs of the present without assaulting the rights of future generations, is addressed and implemented by many countries to manage the environment in an equal manner. However, there are some nations that achieve growth for their economies without regard to the adverse effects on the open environment. Nuclear and related applications became available everywhere to solve many problems of humanity. These applications, if not managed correctly, may lead to adverse effects of contaminating our environment by adding radioactive materials to already existing radioactivity of natural origin. So that our future generations and we will not be the victims of the various contaminations with

IntechOpen

hazards, we must preserve our environment. Many models arise when large companies offer their products without paying attention to long-term effects on human health or environmental stability. Some states allow the export of banned products or remainders inside the country because they are not safe in domestic use. In that regard, there are some talk about agreements to bury dangerous waste (e.g., radioactive waste in deserts). We recognize this by developing appropriate solutions and standards to perform the required tasks. These procedures often require the availability of accurate information and must be much easier to facilitate making decisions. The environmental radiation monitoring, for example, requires a variety of measurements, so it needs development of equipment capable of performing fast and accurate measurements on demand in addition to training of people that deals with radioactive materials.

2. Natural radioactivity

In nature, there are important components that cast a shadow over the existing development of humankind such as uranium, which contributed greatly to the generation of electricity around the world. This element, in addition to other natural radionuclides, believed to be originated during the supernova explosion millions of years ago and/or alien to the earth where it was formed in the fusion of neutron stars, eventually makes its way into the earth crust. Natural radioactivity is a term used to describe the levels of naturally occurring radionuclides in different environmental compartments, originated either from cosmic (e.g., ^{14}C and ^{3}H) or terrestrial radiation. In addition to radioactive potassium (^{40}K), the terrestrial radionuclides include those contained in four known decays series, namely, uranium, thorium, actinium, and neptunium, which start with ^{238}U, ^{232}Th, ^{235}U, and ^{237}Np, respectively. They comprise 18, 11, 16, and 12 radionuclides, respectively. The most abundant in significant levels in our environment are those from ^{238}U and ^{232}Th series. It is believed that in the history of the earth, the crust was enriched in uranium in the beginning; then, the rise of oxygen had oxidized uranium leading to the transfer of huge amount to the oceans and by some natural processes back to the mantle [1, 2]. The processes involved led to spatial distribution of uranium and its decay products. No matter how these theories and assumptions are exact, they give a picture of the approach we can only prove their validity by experiments.

Natural environmental radioactivity arises primarily not only from uranium, as mentioned above, but includes also other nuclides, such as thorium series and potassium, which occur at trace levels in all formations. These radionuclides are believed to be formed by the process of nucleosynthesis in stars and are characterized by half-lives that are comparable to the age of the earth.

It has been recognized that there are some places with large inhabitants that encompass high levels of background radiation in environmental compartments. Great interest given worldwide for the study of naturally occurring radionuclides has led to the performance of broad investigations in many countries [3–10]. Investigators attempted to correlate the distributions of natural radionuclides with some settings such as geology, soil characteristics, etc. Such surveys can be useful for both the assessment of dose rates and the exploit of epidemiological studies, as well as to keep reference-data histories and to determine possible changes in the environmental radioactivity due to nuclear, industrial, or any other practices. What matters to us is to deal with the current reality of taking advantage of natural resources without disruption and tampering with our environment. The accurate determination of isotopes in environmental media presents a significant contest. Thanks to the technology that offered today many nuclear and related techniques for evaluating isotopes in the environment in efficient manner. Depending on the isotope, the analytical technique is selected (alpha, beta, or gamma emitter).

Gamma radiation emitted from naturally occurring radioisotopes, also called terrestrial background radiation, represents the main external source of irradiation of the human body. Natural environmental radioactivity and the associated external exposure due to gamma radiation depend primarily on the geological and geographical conditions, as reported at different levels in the soils of different regions around the world [11–13]. The specific levels of terrestrial environmental radiation are related to the geological composition of each lithologically separated area and to the content in thorium (Th), uranium (U), and potassium (K) of the rock from which the soils originate in each area.

2.1 Radiation protection from natural sources

This topic received some interests by many researchers in the field. Regardless of the general situation of safety and exposures, there are a number of conceptual issues, which remain open. That may include better revision of the protection concepts to cope with conditions of long-term chronic exposure resulting from natural sources. Developing real-world methodologies for the assessment and regulation of situations where there is a potential of exposure and addressing long-term safety aspects of radioactive waste of natural origin deem necessary. For decades, several studies have been conducted on the behavior of radionuclides in the environment and their transfer to humans through ecological and food chains. Most research focused on the contamination of the food chain release to the environment and development of mathematical models to describe environmental transport and assessment of general exposure. Continuing basic biological research is of particular importance to progress in protecting human, animal, and the environment from the hazards of radiation, so it should be strongly supported. However, it is also important to allow epidemiology, especially studies of low-dose populations, and to improve understanding of environmental phenomena as they relate to radiation protection, so as not to throw our hands at risk.

Many practices nowadays may increase the risk of surface contamination by radioactivity which needs control, such as oil exploration leading to NORMs, phosphate fertilizers, and illegal disposal of radioactive wastes in remote areas. Environmental monitoring can afford valuable means for understanding the distribution of natural worries of the ecological system. It is therefore importantly needed to increase our knowledge of the system by better means and offer adequate information to regulators, decision-makers, and the public. Authorities and investigators make baseline data such as risk maps to identify areas with low or high concentrations of certain radioactive and nonradioactive elements.

2.2 Environmental samples and sampling

Environmental sample includes anything on the earth (soil, rocks, plants, water, sediments, air, etc.). It is important that samples taken from any place have to be representative to that place and care necessity be taken not to cross-contaminate samples. These precautions include also storing samples in a safe place to prevent conditions that could change the properties of the sample. Samples shall be kept sealed during long-term storing or transport. Before sampling a protocol, sampling strategy has to be set and all records of field sampling are written in a certain logbook. Simple logbook contains basic information of samples and sampling (date/time, coordinates, climate conditions, dose rate readings, etc.) It may contain additional information such as where and how samples are taken. As an example, soil samples can be taken using auger with depths up to 20 cm (after removing the top 2–3 cm). Locations of samples have to be pre-defined on approximate map, and from each location, a set of triplicate samples (as shown in **Figure 1**) could be taken. Samples are then prepared for measurements in standard procedures (drying, grinding, sieving, etc.). Details about sample preparation are described elsewhere such as the IAEA Technical Report Series No. 295 [14].

2.3 Gamma radiation monitoring

Among other types of radiation, gamma rays are the most penetrating radiation that are emitted from natural and manufactured sources. This property made gamma rays easy to detect and measure. Measurements can be made in two manners: total measurements that record gamma rays emitted at different energies from various sources. These modes are generally used to evaluate the gross levels of the gamma radiation in fields and to detect the presence of abnormalities in the environment. Laboratory analyses, on the other hand, measure both the intensity and energy of radiation, which enables identification of the source of the radiation.

Gamma radiation monitoring is applied in several fields of science including geological, geochemical, and mineral exploration, related epidemiological studies, and environmental science. It allows the interpretation of regional features over large areas. The monitoring is useful to estimate and assess the terrestrial radiation dose to the human population and to recognize regions of probable natural radiation hazard. Radioactive potassium and the uranium and thorium decay series are relatively abundant in the natural environment. They produce gamma rays of sufficient energies and intensities to be detected by a simple gamma ray spectrometry. Average crustal abundances of these elements quoted in the literature are in the range 2–2.5%, 2–3 ppm, and 8–12 ppm for potassium, uranium, and thorium, respectively [12].

Regional monitoring provides a base against which contamination from artificial sources be estimated. For example, regular measurements are conducted around nuclear facilities such as power plants, hospitals, and mining, industrial, and even radiowaste sites to provide a baseline against which any unintentional release of radioactive material can be detected. The gamma ray techniques have been fruitfully applied to mapping the fallout from nuclear accidents [15].

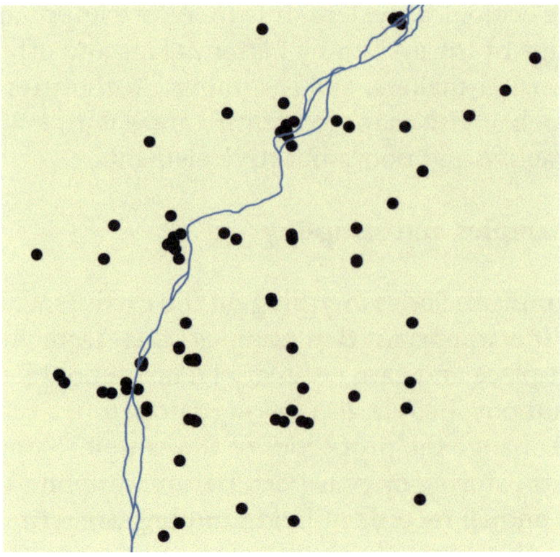

Figure 1.
Part of soil sampling area where triplicate samples are found from each location in the area.

2.4 Gamma ray laboratory

Gamma spectrometry is a system that is equipped with various types of detectors (HPGe, BEGe, LEGe, NaI, etc.), which characterized its specifications for radioactivity measurements. Germanium detectors are powerful systems used to measure the radioactivity in environmental samples. They have many advantages compared to other techniques as, for example, they distinguish many radionuclides in one single measurement without destruction or chemical modification of the sample. Simultaneous identification of many radionuclides with specific gamma energy and high-energy resolution of the germanium detectors allows measurements of complex combinations of gamma emitters. **Figure 2** shows typical gamma spectrum taken for environmental sample.

It is important to know what counting statistics is used to optimize counting times in view of the influence of background. Depending on detector characteristics, the minimum detectable activity (MDA) at specific energy E is an important parameter to be calculated for field measurements; this may be given using Eq. (1):

$$MDA = \frac{\sqrt{R(E)B(E)}}{\varepsilon(E)}, \tag{1}$$

where R(E), B(E), and ε(E) are resolution of the detector (keV), background (counts/keV), and total efficiency at the specific energy E, respectively.

2.5 In situ ambient dose measurement

Measurements are generally carried out using various radiation survey meters that can have different detection abilities. The choice of field measuring devices usually depends on how sensitive these devices are to different energies of different concentrations of radionuclides in the environment. Quality control has to be conducted by researcher and investigators to make sure these devices are reliable, accurate, and precise.

An example of the reliability of field, compared to the laboratory measurements, will be given here. In a recent survey, a portable dose rate meter device (Radiogem2000 with probe [16]) was set to measure dose rates, D_F (μSv/h) at 1 meter above the ground while at the same time taking soil samples from the same

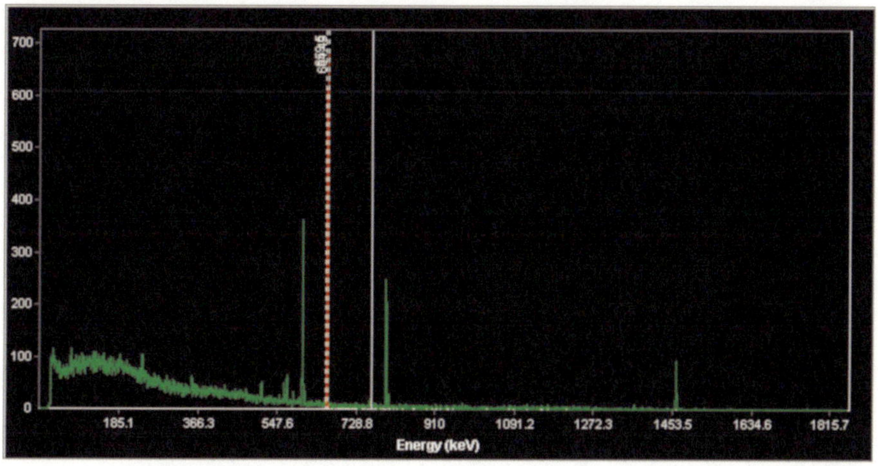

Figure 2.
A typical gamma measurement spectrum obtained using HPGe system.

locations for laboratory analyses of ^{238}U, ^{232}Th, and ^{40}K in the collected samples. Ambient dose rates (D_C) are calculated from the measurements using Eq. (2):

$$D(nGy/h) = 0.4\,61A_U + 0.623A_{Th} + 0.0414A_K \quad (2)$$

where A_U, A_{Th}, and A_K are the activity concentrations (Bq/kg) of ^{238}U, ^{232}Th, and ^{40}K, respectively [11].

As shown in **Figure 3**, a very good linear relationship between field and laboratory measurements (calculated absorbed dose) was clearly perceived for about 100 data points with moderate dose rates ($D_C \approx 0.7D_F$, $R^2 = 0.97$). Of course, this result could be validated with more measurements. The most important outcome of that investigation is that at normal situations where the absorbed dose is up to 300 nGy/h, the field measurements have good agreement with laboratory measurements. It is therefore safe to rely on the portable devices for routine monitoring. The implication of that is that many measurements could be performed in a field mission (as the measurement takes only few minutes long). If levels are high, then sampling and laboratory measurements force itself.

2.6 Real-time radiation monitoring in the environment

The level of background radiation can be used as a consideration in remedial actions if contamination occurs. If measured constantly, it gives info about the trends with time and impact of man-made activities. Hence, it is important to carry out systematic investigations on ambient gamma dose throughout to establish a baseline database for future control assessment where it acts as early warning system.

The early warning system is composed of detectors installed at different locations and connected to central server over available communication system. Any type of detector or survey meters could be installed and used to fulfill the requirements. The advantage of this system is that the authority can create a national radiation map, showing environmental radiation levels (gross count of the radioactivity) throughout certain area updated in real time. It allows the citizen (or anyone) to see what radiation levels are within that specific area at any instance.

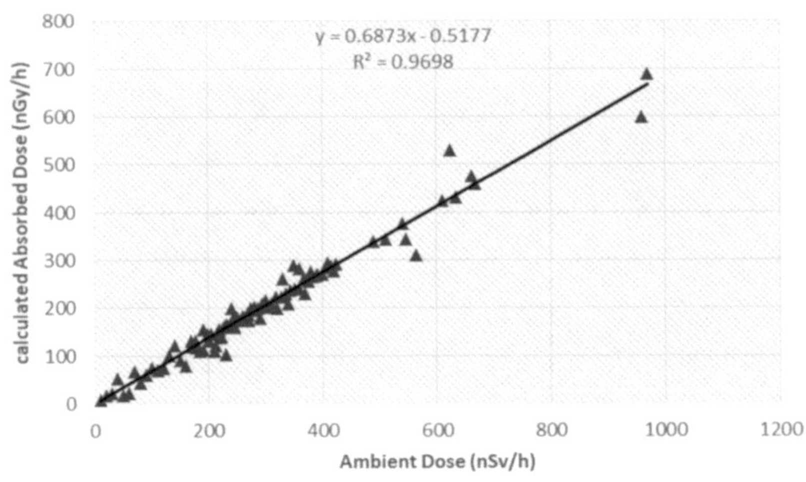

Figure 3.
Relationship between field (ambient dose) and laboratory measurements (absorbed dose).

2.7 Geographical information system (GIS) for radioactivity monitoring

In a simple form, GIS is defined as a set of computer hardware and software designed to acquire, store, manipulate, display, and report geographically referenced information for a particular purpose in space. The space is presented by geographic coordinate systems. Therefore, GIS defines the relationships between various database information and geographical locations within the location system. Together with geostatistical tools, GIS is useful to interpolate scatter data by converting measured points into continuous surfaces. There are several methods available, the choice of which depends on the data itself. Among these methods it is worth to mention two methods, namely, inverse distance weighting (IDW) and kriging.

2.7.1 Inverse distance weighting (IDW)

In this interpolator, the data points are weighted during process so that the impact of points relative to each other is a function of inverse distance. Weighting is calculated to data via the use of a weighting power and the radius object. Larger power means that the adjacent points have the larger influence. Searching radius could be fixed or variable (with typical values of power around two). This flexibility allows controlling the interpolation, which may depend on the number of samples and how they are spatially distributed. One of the drawbacks using this method is that maxima and minima are always among data points since the inverse distance weighted interpolation is a smoothing technique by definition. On the other hand, it is a powerful interpolation technique which leads to reasonable predictions with no problem with results exceeding the range of meaningful values. Simple or advance GIS software could be employed to interpolate and validate the results. Validations are normally expressed as root mean squares error in the correlation between the predicted and actual values.

2.7.2 Kriging method

This is an advance method that makes a surface from scattered points. It is sometimes called weighted moving averaging method because it is derived from regionalized variable theory. It assumes that the variation of a parameter is statistically correlated all over the area. Kriging derives weights from semivariogram functions that depict the degree of spatial correlation between data points as a function of distance and directions between points. The semivariogram adjusts the way kriging weights are allocated to each data point during interpolation. The semivariogram $\gamma(h)$ function is given by Eq. (3):

$$\gamma(h) = \frac{1}{2N(h)} \sum_{i=1}^{N} \sqrt{[Z(x_i + h) - Z(xi)]^2} \tag{3}$$

where $x_i + h$ and x_i are sampling position separated by a vector h, $Z(x_i)$ is a random variable at fixed position x_i, and $N(h)$ is the number of data pairs separated by a vector h. Ordinary kriging is a type of kriging that uses the sampled main variable to estimate values at unsampled locations. Cokriging, on the other hand, allows secondary variables to be incorporated in the model assuming that both primary and secondary variables are correlated [17, 18].

2.7.3 Example of prediction

Figure 4 shows a typical example to predict unsampled places from randomly scattered data points of measured ambient dose (figure to the left). Both interpolator methods IDW and kriging were used to create continuous surface of this parameter as shown in middle and right figures, respectively. These maps (easy to visualize if there are trends) could be used as a guide for any future studies; it can be improved and updated.

2.8 Radon monitoring

Radon is a naturally occurring radionuclide that is found in the environment as a member of the natural decay series of uranium. The ^{222}Rn, including its progeny, is one of the most significant natural sources from a viewpoint of human radiation exposure to the population. Exposure to high concentrations of radon has been correlated to lung cancers, although the effect of low radiation doses is not well defined. The importance of environmental ^{222}Rn data were pointed out in the UNSCEAR reports [11–13]. As an alpha emitter, the indoor ^{222}Rn can be measured using detectors that estimate alpha particles or via its decay products that emit alpha, beta, or gamma rays. Many techniques have been developed to measure radon in the environment. Charcoal canister technique and solid-state nuclear track detector (SSNTD) are common methods in use to evaluate radon in passive mode. The radiation doses due to radon inhalation are calculated according to the ICRP assumption of equilibrium factor (the quotient of the equilibrium equivalent concentration to the ^{222}Rn concentration) of 0.4 and assuming 5700 h spent indoors annually.

2.9 Assessment of external hazards

In addition to absorbed dose calculated from Eq. (2), the following additional hazard index parameters are, generally, evaluated using field or laboratory measurements to assess the risk of exposure due to natural radioactivity.

2.9.1 Annual effective dose (E)

The annual effective dose is a quantity that is introduced in the field of radiation protection for dose limitation, defined as organ or tissue weighted sum of equivalent dose in 1 year (averaged for the whole body) considering type of radiation. It represents the stochastic risk (probability of getting cancer [estimated as 5×10^{-2}

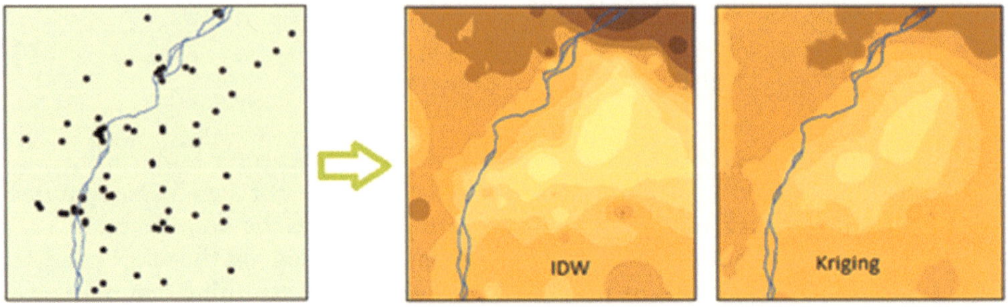

Figure 4.
Converting scattered measured ambient dose to continuous surface using inverse distance weighting (IDW) and kriging geostatistical methods.

per sievert] and genetic defects). For people living in a certain area, the annual effective dose could be calculated using Eq. (4) [12]:

$$E(\text{mSv/y}) = D(\text{nGy/h}) \times 24\text{ h} \times 365.25\text{ d} \times 0.2 \times 0.7\ (\text{Sv/Gy}) \tag{4}$$

where 0.7 is the absorbed/ambient dose conversion factor and 0.2 is the outdoor occupancy.

> Example 1: About 100 soil samples were collected from an area, measured by gamma spectrometry which showed the following average results: 80 ± 7, 91 ± 21, and 573 ± 89 (Bq/kg) for ^{238}U, ^{232}Th, and ^{40}K, respectively. Estimate the annual effective dose for people living in this area spending 60% of their time indoor.
>
> Solution: First we calculate the absorbed dose using Eq. (2):
>
> $$D = 0.461x80 + 0.623x91 + 0.0414x573 = 117\ \text{nGy/h}$$
>
> This can then be converted into annual effective dose using Eq. (4):
>
> $$E \approx 0.3\ \text{mSv/y}$$

2.9.2 External hazard index (hex)

This index is calculated using Eq. (5) [19]:

$$H_{ex} = \frac{A_{Ra}}{370} + \frac{A_{Th}}{259} + \frac{A_K}{4810} \tag{5}$$

> Example 2: In example 1 above, calculate the external hazard index.
>
> Solution:
>
> $$H_{ex} = \frac{80}{370} + \frac{91}{259} + \frac{573}{4810} = 0.69,$$
>
> less than unity (the recommended limit for external exposure).

2.9.3 Excess lifetime cancer risk (ELCR)

Cancer risk can be estimated using Eq. (6) [12, 20]:

$$\text{ELCR} = E \times DL \times RF \tag{6}$$

where DL is the life expectancy (in years) and RF is the cancer risk factor for each sievert [21], which is of order 0.05 for the public.

> Example 3: In example 1 above, estimate cancer risk for a person living in that area.
>
> Solution:
>
> Assuming the average life expectancy of people in this area is 65 years, then using Eq. (6) the lifetime cancer risk is calculated as
>
> $$\text{ELCR} = 0.3 \times 10^{-3} \times 65 \times 5 \times 10^{-2} = 9.8 \times 10^{-4} \approx 10^{-3}$$

3. Conclusion

The chapter describes the importance of radioactivity monitoring to preserve our environment. It sheds the light on methods designated for the measurement of natural radionuclides in the environment and assessment of radiation exposure to human in different situations. In addition to measurements and surveys, the chapter presents a summary of some methods of radiation dose calculations that the individual may be exposed to. As is difficult to measure everywhere, the chapter also presents methods for estimating and predicting the spatial distribution of radiological quantities. The use of a geographical information system, GIS, and geostatistical methods to create maps facilitates the evaluation and assessment of radioactivity in the environment. Environmental measurements may be costly and time-consuming practices; hence, thoughts to reduce time and efforts are given in this chapter where at normal levels portable simple equipment proved useful.

Acknowledgements

Part of the data presented in this chapter is prepared with the support of the "Environmental group of Sudan Atomic Energy Commission." The author would like to thank all members of this group for their collaboration during field missions, laboratory analyses, and reporting.

Conflict of interest

The author discloses no potential conflicts of interest.

Author details

Isam Salih Mohamed Musa[1,2*]

1 Physics Department, College of Science, Taibah University, Medina, Saudi Arabia

2 Radiation Safety Institute, Atomic Energy Commission, Khartoum, Sudan

*Address all correspondence to: isamsalih@gmail.com

References

[1] Andersen MB, Elliott T, Freymuth H, Sims KWW, Niu Y, Kelley KA. The terrestrial uranium isotope cycle. Nature. 2015;**517**:356-359

[2] Collerson KD, Kamber BS. Evolution of the continents and the atmosphere inferred from Th-U-Nb systematics of the depleted mantle. Science. 1999;**283**: 1519-1522

[3] Abdalhamid S, Salih I, Idriss H. Gamma absorbed radiation dose in Marrah mountain series, western Sudan. Environment and Earth Science. 2017; **76**:672

[4] Malczewski D, Teper L, Dorda J. Assessment of natural and anthropogenic radioactivity levels in rocks and soils in the environs of Swieradow Zdrojin Sudetes, Poland, by in situ gamma-ray spectrometry. Journal of Environmental Radioactivity. 2004; **73**:233-245

[5] Sohrabi M. World high background natural radiation areas: Need to protect public from radiation exposure. Radiation Measurements. 2013;**50**: 166-171

[6] Rafique M, Khan AR, Jabbar A, Rahman SU, Kazmi SJA, Nasir T, et al. Evaluation of radiation dose due to naturally occurring radionuclides in rock samples of different origins collected from Azad Kashmir. Russian Geology and Geophysics. 2014;**55**(9): 1103-1112

[7] Al-Sulaiti H, Nasir T, AlMugren KS, Alkhomashi N, Al-Dahan N, Al-Dosari M, et al. Determination of the natural radioactivity levels in north west of Dukhan, Qatar using high-resolution gamma-ray spectrometry. Applied Radiation and Isotopes. 2012;**70**: 1344-1350

[8] Banzi FP, Msaki P, Makundi IN. A survey of background radiation dose rates and radioactivity in Tanzania. Health Physics. 2002;**82**(1):80-86

[9] Beamish D. Environmental radioactivity in the UK: The airborne geophysical view of dose rate estimates. Journal of Environmental Radioactivity. 2014;**138**:249-263

[10] Jeevarenuka K, Sankaran PG, Hameed PS, Mathiyarasu R. Evaluation of natural gamma radiation and absorbed gamma dose in soil and rocks of Perambalur district (Tamil Nadu, India). Journal of Radioanalytical and Nuclear Chemistry. 2014;**302**:245-252

[11] UNSCEAR. Sources and Effects of Ionizing Radiation. New York: UNSCEAR; 1993

[12] UNSCEAR. Effects of Atomic Radiation to the General Assembly. New York: United Nations scientific committee on the effect of atomic radiation; 2000

[13] UNSCEAR (United Nations Scientific Committee on the Effect of Atomic Radiation). Sources and Effects of Ionizing Radiation. New York, United Nation: Report to the General Assembly; 2008

[14] IAEA. Measurement of Radionuclides in Food and the Environment, a Guidebook. Vienna: International Atomic Energy Agency; 1989. Technical Report Series No. 295

[15] IAEA-TECDOC-1363. Guidelines for radioelement mapping using gamma ray spectrometry data; 2003

[16] Available from: https://www.pinterest.com/pin/186406872046885328/

[17] Salih I, Pettersson HBL, Lund E, Ake S. Spatial correlation between radon (^{222}Rn) in groundwater and bedrock uranium (238U): GIS and geostatistical

analyses. Journal of Spatial Hydrology. 2002;**2**(2):1-10

[18] Salih I. Radon in Natural Waters, Analytical Methods; Correlation to Environmental Parameters; Radiation Dose Estimation; and GIS Applications. PhD thesis. Sweden: Linkoping University; 2003. ISBN: 91-7373-510-8; ISSN: 0345-0082

[19] Gulan L, Milenkovic B, Zeremski T, Milic G, Vuckovic B. Persistent organic pollutants, heavy metals and radioactivity in the urban soil of Priština City, Kosovo and Metohija. Chemosphere. 2017;**171**:415-426

[20] Chandrasekaran A, Ravisankar R, Senthilkumar G, Thillaivelavan K, Dhinakaran B, Vijayagopal P, et al. Spatial distribution and lifetime cancer risk due to gamma radioactivity in Yelagiri Hills, Tamilnadu, India. Egyptian Journal of Basic and Applied Sciences. 2014;**1**(1):38-48

[21] International Commission on Radiological Protection ICRP. Recommendations of the ICRP. New York: Pergamon Press; 1990. ICRP Pub No. 60